The Physics of Climate Change

Lawrence M. Krauss

Post Hill
PRESS

A Post Hill Press Book
ISBN: 979-8-88845-092-5

The Physics of Climate Change
© 2021 by Lawrence M. Krauss
All rights reserved
First Post Hill Press Hardcover Edition: March 2021

Cover design by M. Simonsen
Author photo by Nancy Dahl

No part of this book may be reproduced, stored in a retrieval system, or transmitted by any means without the written permission of the author and publisher.

Post Hill Press
New York • Nashville
posthillpress.com

Published in the United States of America
1 2 3 4 5 6 7 8 9 10

For Woody,
who convinced me to keep writing

No man ever steps in the same river twice,
for it's not the same river, and he's not the same man.

—Heraclitus

TABLE OF CONTENTS

PREFACE

Plus ça change, plus c'est la même chose
—Jean-Baptist Alphonse Karr, 1849

It has been almost two years since *The Physics of Climate Change* first appeared in hardcover, and with the release of this paperback edition, it seemed appropriate to add some comments here to discuss whether any significant developments have taken place that might make my earlier discussion outdated. The short answer is no.

After all, this book is about the underlying physical processes associated with climate change, and the major empirical trends, including global warming, sea level rise, and glacial melting that have been associated with the increased CO_2 content in our atmosphere due to human industrial activity over the past 150 years. As a result, the fundamental issues I have discussed remain as current today as they were two years ago.

One might say that, if anything, they have become more pressing, because while there was a slight dip in global atmospheric CO_2 production during the early stages of the pandemic during which I wrote the book, there has been a return to business as usual, and the 2021 global emissions from fossil fuel burning, at 37.08 billion tons, exceeded the pre-pandemic 2018 value of 36.83 billion tons, and the increase in atmospheric CO2 content during 2021, of 2.58 parts per million (ppm) over 2021 was tied for the 5th highest annual increase in the National Oceanographic and Atmospheric Administration's sixty-three year record. By January 2023, the CO_2 concentration had reached

419.62 ppm, a further 2.51 ppm increase from a year earlier. The march upward continues with, as of yet, no signs of decline. At the same time, ice sheet melting in Greenland has continued with a loss of 146 ± 64 billion tons of ice in the 2021–2022 melting season, the 25th consecutive year of ice loss. In Antarctica, an ice shelf on the colder Eastern coast collapsed in 2022, and new December 2021 data on the Thwaites Glacier, which protects the West Antarctic ice sheet, and itself holds enough ice to potentially raise global sea levels by two feet, displayed huge cracks stretch across much of the Glacier.

Nevertheless, these quantitative developments do not qualitatively change any of the fundamental issues I addressed in the hardcover edition of this book in 2021. They are consistent with what was ongoing at that time, and their possible implications were also discussed.

On the theoretical side, some positive news did appear, suggesting the possibility that some of the IPCC models of temperature rise going into their Sixth Assessment Report in 2021 had systematically predicted somewhat more warming over previous periods than was observed, and therefore may predict more warming in the future.

But it must be emphasized that this result does not qualitatively impact on either the overall impact of CO_2 on global temperatures, or on the likely impacts of global warming in this century. And, as climate scientist Tim Palmer has emphasized, because the ensemble of IPCC predicted temperatures has a distribution which is skewed toward higher temperatures, the "most likely" prediction can actually be below the predicted mean temperature rise, by as much as a degree Celsius.

Concerns about global climate tipping points continue and the IPCC and others have issued more urgent concerns about the impacts of a future in which "business as usual" continues through much of this century. And because, as I have described in this book, much of the CO_2 that is emitted into the atmosphere stays in the atmosphere for perhaps 1,000 years, every year that we continue our emissions makes it more difficult to keep the total CO_2 atmospheric content below a value that might keep global temperatures from rising by more than two degrees Celsius. Over the two years since this book first appeared, humanity has dumped an addition 20 billion tons of carbon into the atmosphere.

As I note at the end of this book, the future is rushing at us like a freight train, but it is doing so on tracks we have built. *"Nothing is written"* as T.E. Lawrence exclaimed in the film *Lawrence of Arabia*. We can change the future. I recently visited a commercial carbon capture facility in Iceland that has demonstrated a proof in principle that it can effectively capture and sequester carbon deep underground in rocks. The levels at which it can function are orders of magnitude below what is necessary counter to any great degree the amount of CO_2 being emitted today, but these are early times, and I retain my optimism about technological progress.

The greatest challenge of dealing with global climate change is, however, not a technical one, but a political one. And the most dramatic impacts of climate change may not be physical ones, but rather political ones as well. Whether humanity has the will, and whether political and business leaders can muster the courage and the support to address these issues in a way that enhances our global quality of life remains to be seen. But, as I indicate in the foreword of this book, the first step on deciding whether we should act is to understand the underlying science, and to be aware of the empirical date. That was my original intent, and it remains true today.

Climate science is no different than other science, and as I demonstrate, it is based, to large degree, on well-established principles developed and tested over a century ago. The story of how we have come to know what we know remains as interesting today, independent of the urgency of these issues, as it ever was. I do not advocate here for specific policies, on the right, or the left. Science, and scientists, should provide advice on how the world works, and what is likely and what is not. It remains for the rest of us to decide how we use that information.

LAWRENCE M. KRAUSS, PEI JANUARY 2023.

FOREWORD

1

EARLY ONE AFTERNOON IN JANUARY OF 2020, I FOUND MYSELF sitting alone at the bow of a riverboat traveling down the Mekong River from Phnom Penh to Ho Chi Minh City. I was finishing preparations for a lecture and enjoying the sunshine and breeze while watching the busy river traffic. Everywhere, there were barges relentlessly digging up sand from the river bottom for later use, among other things, as concrete for building products. According to the Mekong River Commission, sand mining has caused the riverbed to lose 1.4 meters of elevation since 2008.

As I looked around, I began to feel a growing sense of sadness combined with loneliness. Sadness because the lecture I had just finished preparing for travelers on the boat involved the nature and physics of climate change—with a focus on the potential impact for the Mekong Delta. During my research, I had come to realize how a confluence of factors made this region, home to sixty million people—at least fourteen million of whom depend directly on the health of the Mekong Delta—the epicenter of a Perfect Storm, where even the more conservative predictions of global climate change in the next thirty years may devastate the entire area and the lives of the people who live in it.

Many of my fellow shipmates, a few of whom joined me up front as the afternoon wore on, were as of yet unaware of the fragility of the landscape that then surrounded us, and I wasn't eager to burst their bubble later that evening.

After the discussions following my lecture a few hours later, it became clear that while some of the realities were unpleasant, the well-meaning and interested laypeople who had gathered on the boat wanted information to put this global existential issue in perspective. They wanted to figure out how to separate the wheat from the chaff, to see what was at stake, and learn what possible future impacts humanity might and might not be able to affect. That was when I decided to write this book, and I thank my shipmates for inspiring me.

I am not a climate scientist. You may wonder why a particle physicist and cosmologist would wade, literally, into this subject. Because others, whose future depends on the policies governments enact and who also have to assess the discrepant claims emanating from politicians and the media, are not climate scientists either. If it isn't possible to explain the scientific principles and predictions associated with climate change in a straightforward and accessible fashion, then what hope is there for any rational public discourse and decision-making on the subject?

If the goal is to create something that provides readers a reasonably informed perspective of this subject in particular, where does one begin?

First off, it is worth recognizing that climate change science is *not* rocket science. Having once written a book about rocket science, or at least imaginary rocket science, I decided I was in a good position to judge. And the urgency of the issue is surely greater than pondering the possibilities of space travel in the twenty-third century, as fascinating as those might be.

Next, the details of large-scale supercomputer climate models that make detailed predictions about the future are complex and intimidating, but the underlying physics governing global warming is nevertheless straightforward and grounded in basic science. As a plus, it turns out there are historical twists and new connections between scientific disciplines that add spice. And for those who are particularly interested, a wealth of data is now freely available for anyone to follow up with on the web.

⸻

I am fortunate to have been educated by a number of climate change experts who are both colleagues and friends. For over a decade, I was chairman of the Board of Sponsors of the Bulletin of the Atomic Scientists. When I joined the board in 2006, we chose to include climate change as an additional existential threat when we decided upon the setting of the famous Doomsday Clock. Each fall, we would host a symposium to discuss scientific and technological challenges, then the science and security board, which included various climate experts, would further discuss the issues raised during the symposia when we decided on how to set the Clock. Later, I was fortunate to host several scientific meetings and public events on climate change. Most recently, the Origins Project Foundation I lead organized the Mekong cruise I lectured on.

I am thankful for the discussions I had with colleagues during these years, including James Hansen, Richard Somerville, Susan Solomon, Dan Schrag, Tony Haymet, Raymond Pierrehumbert, and the late Wallace Broecker, among others, many of whom also provided me with useful data and figures. I thank these individuals for their intellectual and personal generosity.

Numerous friends, colleagues, and experts were also kind enough to review this book at various stages. I am deeply indebted in particular to Richard Dawkins, Dan Schrag, Penn Jillette, Richard Somerville, Neil deGrasse Tyson, William Frucht, Sheldon Glashow, Keith Ogorek, and John Dahl for critically reading, commenting on, and improving this manuscript. I am also indebted to the numerous scientists who provided me with permission to reproduce figures from their work in this book. Any errors that remain are, of course, my own.

The support and encouragement I received from a host of people during and after the writing of this book have been particularly important. I was surprised and dismayed as numerous publishers and editors I reached out to indicated to me that they thought the only marketable books on climate change would be ones that appeal to emotions and communicate only to the true believers through a sense of doom and gloom. Since they are in some sense the gatekeepers for what information the public gets, this demonstrated to me how important it is to combat that perception with a book that could provide actual information the public can use to make informed decisions about how to respond to what they might read in the papers or hear from politicians.

FOREWORD

The science behind climate change is accessible and interesting, and it should be the basis of arguments and policy discussions. Appealing purely to emotion or using scare tactics should not be the way to encourage action, just as encouraging inaction by denying the evidence and underlying science is inappropriate.

When I reached out beyond the publishing community to friends, colleagues, and fans of earlier books, I was encouraged to find that a book of this character was just what many people felt was needed, and that it should be distributed broadly. I thank all of those who helped reinforce my conviction that this book was necessary and who helped energize my efforts to make sure it reaches people who may find it useful for themselves or in their discussions with others. In particular, Susan Rabiner, Jahm Najafi, Thomas Houlon, Patty Barnes, Marylee MacDonald, Pamela Paresky, and Richard Dawkins all helped me explore a variety of publishing options in my efforts to ensure that this book ultimately reached readers in its present form.

Happily, at the end of this process, I found the marvelous editor and publisher, Adam Bellow. From our very first discussion, it was clear we shared the same vision for the book and the need to ensure that science and reason and free and open dialogue remain an important part of the social fabric. I am very happy this book found a receptive home through Adam at Post Hill Press.

Climate change, evolution, and the Big Bang are all empirical facts, not speculation, and the relevant data validate fundamental theoretical expectations. This convergence reflects science at its best and most powerful. And it is the science I will concentrate on in this book. I will not advocate for specific policies;

that is the purview of politicians, advocacy groups, and political movements. I will, however, be unabashed about the seriousness of the challenges we now face so the risks and possible consequences of inaction are manifest.

It would be disingenuous to imply that my agenda, while primarily scientific, did not also have an associated political purpose. But it is not one characterized by terms like liberal or conservative. It is simply this: Climate policy will ultimately be determined by various competing interests. Whether these reflect the broader interests of the public at large, whose lives, after all, will be most affected, is not obvious. In this, as in all things, governments usually follow rather than lead. Events of the past several decades, reaching a particular crescendo in the past four years, have validated the fact that democracy depends on an informed electorate, as well as informed legislators, if it is to function effectively.

It is in large measure our choice, which of the possible futures afforded to us will be that experienced by our children and grandchildren. We should enter into that future with our eyes wide open.

CHAPTER 1

A RIVER LIKE NO OTHER

I thought how lovely and how strange a river is.
A river is a river, always there, and yet the water flowing
through it is never the same water and is never still.
It's always changing and is always on the move.
And over time the river itself changes too.

—AIDAN CHAMBERS, *This Is All*

TO TRAVEL DOWN THE MEKONG RIVER NEAR ITS DELTA IS TO experience a waterway like no other in the world. Unnavigable for much of its 2,800-mile length, it is the longest river in Southeast Asia and the twelfth-longest river in the world. By the time the river flows past Phnom Penh in Cambodia and onward into Vietnam, the low-lying and flat terrain causes it to spread out into many separate branches. The Mekong basin covers an area the size of France and Germany combined. On average, the river is almost a mile wide and is far wider at many points. Khone Falls, on the Laos-Cambodian border, is the widest waterfall in the world. Its series of rapids and falls are almost seven miles wide, with a drop of seventy feet! All told, the river disgorges over 475 billion cubic meters of water each year into the sea, and it provides food and water for sixty million people.

━━━━━━━

The transformation of the river near the end of its journey to the sea is beautifully captured in John Keay's masterful book *Mad about the Mekong*, which retraces one of the nineteenth century's most remarkable and harrowing triumphs of explora-

1

tion. In 1866, the French Mekong River Commission of twenty men embarked on a two-year journey up the Mekong, traveling a distance longer than the entire length of Africa, to map the full system. They started in Saigon and ultimately made it all the way to the Yangtze in China. Thirteen men survived the journey.

Keay poetically describes the Mekong's final push to the sea:

> The Mekong falls only six meters in its last eight hundred kilometers, but so low-lying is the Delta that the river in flood appears, and often is, the highest thing around. The land is so flat that from an upper deck you must allow for the curvature of the Earth's surface in counting the tiers of a distant pagoda... After forcing its way for thousands of kilometers through mountain gorge and deepest forest, it is as if the river can scarcely believe its good fortune. Like a sluice released, it wells across the plain, exploring the arroyos, tugging at pontoons, basking in backwaters and generally making the most of its first and last unimpeded kilometers...The Delta is said to produce more rice than any area of comparable size in the world. Beneath the glinting panes of water lie meadow and mud at no great depth. But rice-growing being a form of hydroponics, for the last six months of the year the fields are lakes and the landscape is a waterscape.

As lyrical as this description is, it still misses several of the river's most unique features. Because the lower Mekong Delta south of Phnom Penh lies just above sea level and is exceedingly flat, the shallowness of the river produces a striking annual variation. At Phnom Penh, the river is joined by the Tonlé Sap river and lake system. Depending on the season and the river's varying height, the direction of the Tonlé Sap actually changes. At times, it becomes a tributary, flowing into the Mekong. During flood season, the flow reverses, and the floodwaters flow along the Tonlé Sap into its large lake.

Beyond its annual ebbs and flows, the Mekong Delta experiences a daily surge that, while not unique in the world, is nevertheless rare enough to have aroused the fascination of the earliest Western visitors—and surprise all the rest of us who first learn about it. For much of the year, the delta experiences only one high tide from the surrounding China Sea each day.

While former Fox TV host Bill O'Reilly notoriously claimed that no one knows why the tides happen, in fact none other than Sir Isaac Newton explained the fundamental physics behind the tides when he developed his law of universal gravitation in the seventeenth century. As Newton described it, the gravitational force of the moon on the Earth varies as the inverse square of the distance between the moon and the Earth. Therefore, the side of the Earth nearest the moon, being slightly closer to the moon than the center of the Earth, is pulled with a stronger gravitational force than the average force on the Earth. Similarly, the side of the Earth opposite the position of the moon is pulled toward the moon with a smaller force. Ignoring for a moment the motion of the moon around the Earth, as the Earth rotates, one would expect two bulges to occur in the world's oceans: one on the side facing the moon and another on the opposite side. Roughly speaking, the first occurs because the water is pulled away from the Earth, and the second because the Earth is pulled away from the water. Looking down on the Earth-moon system, one would expect schematically to see something like Figure 1.1.

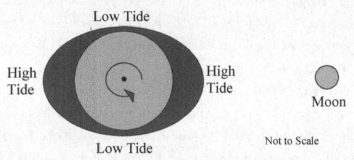

Figure 1.1[2]

As the Earth rotates fully once each day, each spot on the Earth would therefore be expected to experience two high tides and two low tides, as Newton famously described.

In practice, understanding the tides is more complicated. Water on Earth cannot instantaneously relocate to the equilibrium configuration shown in the figure, but must move from place to place, and flow rates depend on local conditions, including ocean depth. The rotation of the Earth and the motion of the moon must also be considered, as must the position of the sun.

Happily, however, we don't need to take all that into account to understand how two tides a day can turn into one tide a day. The key is to recognize that the moon does not orbit around the Earth's equator as the Earth rotates each day. The plane of its orbit is tilted relative to the Earth's axis, varying between eighteen and twenty-eight degrees relative to the equatorial axis of the Earth over an eighteen-year period. Consider the Earth-moon system in a frame where the Earth's axis of rotation is vertical (Figure 1.2).

The tidal response of the ocean, using the reasoning from the figure shown earlier, is seen in Figure 1.3.

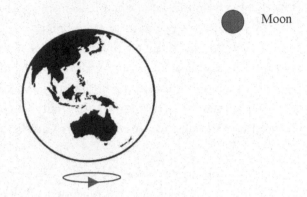

Figure 1.2[3]

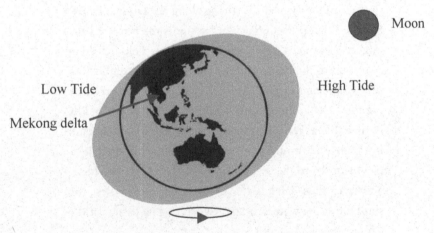

Figure 1.3[4]

The actual magnitude of tides and the relative size of the tides depend on local conditions. But roughly speaking, given the latitude of Vietnam and the position of the moon, as the Mekong Delta rotates around with the Earth, the seas in the region will tend to experience a high tide when southern Vietnam is on the side of the Earth facing the moon, where the bulge is big, and a low tide twelve hours later when it is on the far side, where the water has pulled away and there is no bulge.

I have belabored this issue not just because it involves a bit of astrophysics and is a frequent source of confusion, but because it plays an important role in the future of the Mekong ecosystem that will be relevant later. A large single daily tide impacts on the flow of the Mekong River itself. Again, to quote Keay:

> This diurnal mother-of-a-tide ought, of course, to spell disaster to the Delta. A salty inundation, albeit only once a day, would soon sour the world's most productive rice-bowl and turn the green dazzle of paddy into maudlin thickets of mangroves like those along the Donnai below Saigon. What prevents such a disaster is the power of the mighty Mekong. The inrushing tide meets the outrushing river, and in the best traditions of ecological equilibrium they compromise. The river rises, its progress barred by the tide. The backing-up of the river by a big 'diurnal' is measurable as far upstream as Phnom Penh and beyond. But there and throughout the three to four hundred kilometers down to the sea, salination is barely detectable... The river thus defends the Delta from its deadliest foe since the rising waters are overwhelming its own, not the China Sea's.

The Mekong has the richest density of freshwater fish in the world and is home to what is estimated to be over one thousand species of fish. Directly supporting a population of over fourteen million people, a greater population of freshwater fish is harvested from the Mekong each year than from all the lakes and rivers in the US combined. Over the course of the year, its floods bring water and silt to nourish rice paddies, making the Mekong delta the world's most productive rice bowl.

While some of the most dramatic potential global impacts of climate change might not become manifest for many decades, centuries, or even millennia, the Mekong may be one of the first casualties in the battle to head off Earth 2.0. The shallowness of the river, the flatness of the delta, and the flooding caused both by the seasonal weather and the delicate balance of tides and river dynamics make the Mekong Delta particularly sensitive to even small near-term changes in any of these systems.

It is not just the highly publicized dire predictions of climate change that can have a dramatic impact on many people's lives. The Mekong Delta is a canary in a coal mine for climate change, and that is one reason I have begun this book by describing it and why I shall return to discuss the specific predictions and impacts for the Mekong region at the end of this book. But more than this, precisely because of its unique character, its richness, and its direct impact on a large surrounding population, the demise of the Mekong would have an impact far beyond the confines of Southeast Asia.

While the particular circumstances of the Mekong are unique, various other locations around the world live with similar fragile balances of opposing ecological forces, from the lowlands of Bangladesh to the everglades of Florida and the mouth of the mighty Mississippi. Climate change as a global issue may manifest itself in a thousand different ways in a thousand different places. But just as no man is an island, in an interconnected world, no single place and no single country is likely to be completely immune from the impact of even seemingly small changes first detected on the other side of the planet.

CHAPTER 2

HISTORY AND NUMBERS: HALF EMPTY OR HALF FULL?

Some people see the glass half empty,
some see it half full. I always saw the coffin half full.

— WOODY ALLEN, *Apropos of Nothing*

IF YOU OBSERVE SOMETHING YOU HAVE NEVER SEEN BEFORE, how do you determine if it is good or bad? Or if it is anomalously big or small? Dangerous or benign?

Perspective is everything, of course, but how can we gain it? The problem is numbers alone can be deceiving. The same data viewed in different contexts can appear to present a vastly different picture. Is the glass half-full or half-empty? This was the problem facing those who were first presented with estimates of carbon dioxide (CO_2) concentrations in the Earth's atmosphere.

In 1953, Charles David Keeling had just begun a postdoctoral position at Caltech. Originally he was working on a project to extract uranium from granite, but then moved to another geochemical problem: investigating how carbonate achieves equilibrium in a mixture of water, limestone, and atmospheric CO_2. To do this, he needed to construct a precision device to measure CO_2 extracted from the air and water.

To test his apparatus, he measured the CO_2 concentration in various locations in Pasadena but found significant variations, which he figured were probably due to locations of heavy industry. So he took the equipment up to a more isolated spot: Big Sur, near Monterey, California. During every afternoon he measured the same value of CO_2 concentration in the atmosphere, 310 parts per million (ppm). He started taking samples during both night and day to get better estimates and discovered a diurnal pattern he hadn't anticipated. At night, there was more CO_2 in the air than there was during the day. Also, with great prescience, he measured the ratio of ^{13}C to ^{12}C and discovered this ratio was smaller at night than during the day.

Checking a meteorology book, Keeling discovered that the concentrations he measured were representative of uniform mixing with the "free-atmospheric" concentrations that prevailed over the continent. However, at night there was a lower boundary layer so the air measured was more heavily influenced by concentrations near the ground, where local plant and soil respiration would be systematically more important. This interpretation was confirmed by a decrease in the $^{13}C/^{12}C$ ratios at night, as plants preferentially respire ^{12}C.

In 1956 Roger Revelle at Scripps Institution of Oceanography and Harry Wexler at the US Weather Bureau joined with Keeling to suggest a bolder global measurement of CO_2 during the upcoming International Geophysical Year (1957–58) at a variety of remote locations presumably unaffected by local contamination, including the South Pole station and at Mauna Loa in Hawaii.

In March 1958 Keeling installed his first infrared gas analyzer at Mauna Loa, and on its first day of use, it recorded a concentration of 313 ppm. This was the first reading in what has become one of the most significant continuous terrestrial scientific projects ever carried out. It has been ongoing for the last sixty-two years and has given the world its first quantitative assessment of the impact of global industrial activity on the composition of the atmosphere.

Over the first few months of its installation, Keeling noticed a surprising monthly increase in CO_2 concentrations up until May. After this it declined again until October, and this pattern repeated in 1959. It was as if the Earth was breathing in and out once each year.

It actually is, due to the existence of life on our planet.

Life has changed the composition of the Earth's atmosphere since its inception some four billion years ago, and it continues to govern the dynamics of CO_2 on the planet today. What Keeling was detecting was a modern annual cyclic version of the ancient processes of life that first produced the existing atmosphere on the planet today. He was directly observing, for the first time, the seasonal impact life has on the atmosphere in the Northern Hemisphere through the process of photosynthesis in plants, which converts CO_2 and water into organic compounds with O_2 as a residue. As Keeling later put it in a 1960 article, "We were witnessing for the first time nature's withdrawing CO_2 from the air for plant growth during summer and returning it each succeeding winter."

The second significant observation Keeling made between 1958 and 1960 at Mauna Loa was that the average concentration of CO_2 measurably increased during this period. If one considers comparable months each year, March CO_2 levels went from 313.4 ppm in 1958 to 314.4 ppm in 1960. A somewhat more dramatic effect was observed from samples taken from surface flasks collected at the South Pole, which increased from 311.1 ppm in September 1957 to 314 ppm in September 1959.

Were such small increases significant? Noting that the increase observed at the South Pole was consistent with what one would expect from the terrestrial combustion of fossil fuels, Keeling was nevertheless well aware that claiming a trend based on such a small time sequence was dangerous. Here is how he put it in his first paper for the Scripps Institution of Oceanography in March of 1960:

> Where data extend beyond one year, averages for the second year are higher than for the first year. At the South Pole, where the longest record exists, the concentration has increased at the rate of about 1.3 ppm per year. Over the northern Pacific Ocean, the increase appears to be between 0.5 and 1.2 ppm per year. Since measurements are still in progress, more reliable estimates of annual increase should be available in the future. At the South Pole the observed rate of increase is nearly that to be expected from the combustion of fossil fuel (1.4 ppm), if no removal from the atmosphere takes place. From this agreement one might be led to conclude that the oceans have been without effect in reducing the annual increase in concentration resulting from the combustion of fossil fuel. Since the seasonal variation in concentration observed in the northern hemisphere is several times larger than the annual increase, it is reasonable to suppose, however, that a small change in the factors producing this seasonal variation may also have produced an annual change counteracting an oceanic effect.

This skepticism is the hallmark of good science. Was the observed increase significant? If the expected annual output of CO_2 due to human industrial activity was smaller in magnitude than the seasonal variation itself, was it possible to claim a signal amidst the noise? What was the role of the oceans in possibly moderating this signal? Beyond this, did a CO_2 concentration of 310 ppm itself reflect anything important about the global dynamics of the planet as it is affected by life, including human life?

These were all important questions in 1960, some of which had been anticipated a few years earlier by the pioneering oceanographer Roger Revelle and his colleagues at Scripps. This is probably one of the reasons Revelle enlisted Keeling in his own effort. To begin to answer these questions, accurate measuring would have to be carried out from remote locations regularly for long periods. That, of course, is what has happened. Every day for the past sixty-two years at Mauna Loa, CO_2 has been measured by the technique first used by Keeling, involving infrared gas analyzers that measure the absorption of infrared radiation in atmospheric samples and compare it with absorption rates in samples with known calibrated CO_2 concentrations. The result has been one of the most famous plots in science, appropriately named the Keeling Curve. Scripps Institution updates the curve every day and presents the data for public consumption. Figure 2.1 shows the curve up to the day I wrote this chapter, when the CO_2 reading was 415.19 ppm.

The seasonal variation first observed by Keeling is obvious, but now so is the monotonic rise in the average value of the CO_2 concentration year over year that was first tentatively inferred by Keeling in 1960. Recall that, in 1958, the peak abundance was about 315 ppm. The value today is therefore more than 30 percent larger than the measured atmospheric abundance at that time.

What are we to make of this? As a physicist, the first thing I generally do is examine orders of magnitude, which usually give some perspective on any measurement. The CO_2 abundance has increased by about 100 ppm in sixty-two years. This is about 1.6 ppm per year. Recall that Keeling himself estimated the CO_2 generation by fossil fuel consumption in 1960 to be about 1.4 ppm, about the same order of magnitude.

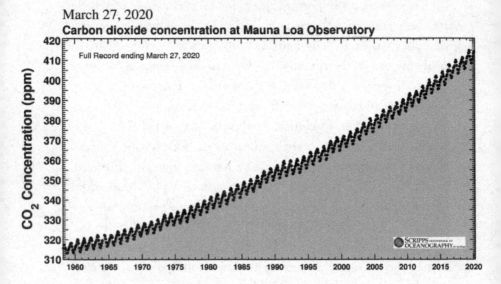

Figure 2.1[5]

Note, however, that global annual CO_2 emissions have increased by a factor of five since 1960, while the slope of the Keeling curve has not increased by a similar factor. Should this give us pause? Perhaps, except Keeling himself pointed out a very plausible reason why one might expect that not all the CO_2 being produced by humanity would be reflected directly by a concomitant rise in atmospheric CO_2 content. CO_2 can dissolve in water, creating carbonic acid, so one would expect some CO_2 to be taken up by the oceans. So, the rough quantitative correlation between the observed rise and our human contribution is, at the very least, suggestive.

Perspective is everything, however, and we must always remember a key warning in science: correlation does *not* imply causation. So, it is possible, without some underlying physical explanation and without more data, that the comparable rates of CO_2 in the atmosphere and human-generated CO_2 is just a coincidence.

To get a better idea of whether the observed CO_2 increase is correlated to human industrial activity, we can hope to explore longer-term variations to see if the current fossil fuel-generating era is anomalous or not. But Charles Keeling wasn't making direct measurements before 1958, and before his time the few direct measurements that were made were scattershot and discordant.

Fortunately, however, nature has given us a time capsule. In places that remain frozen all year long, ice builds up as snow falls. It is for this reason that places like Greenland and Antarctica have ice sheets over a mile thick. Like the growth of tree rings or sedimentary layers of rock, as one drills deeper into ice, one encounters ice layers that were deposited at ever-earlier times. And like growth rings, ice deposition is different in summer than in winter, so a regular pattern allows one to literally count years. Many cores are extracted from each area so better estimates can be obtained.

Figure 2.2 shows examples of core sites from a recent US International Trans-Antarctic Scientific Expedition study in Antarctica.

In the ice there are bubbles. Air gets trapped in the snow as it is compressed, and the bubbles therefore reflect the atmosphere at the time and place when the ice first formed. Measure the gas composition of the bubbles and you know the gas composition at that time and place. Since Greenland and Antarctica have the deepest, largest, and historically the most stable ice conglomerations, most ice cores come from these two locations, and at their upper layers, can be compared with the direct measurements going back to 1958 to calibrate the ice core estimates. Because each location gets a lot of snow each year, ice from successive years can be visually separated relatively easily so good time resolution is possible as well. It is also fortunate that Greenland

and Antarctica are in opposite hemispheres, so in some sense comparing values from ice cores in both locations gives us both a global consistency check as well as a global average for each period.

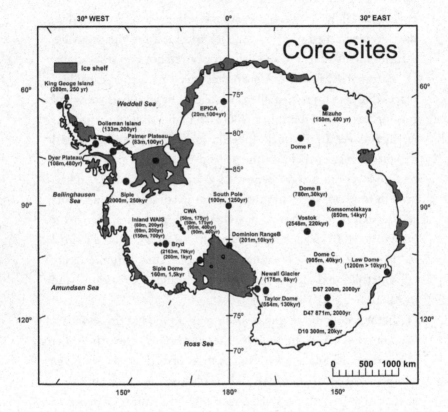

Figure 2.2[6]

Let's take a walk back in time using the ice core data. Scripps Institution provides graphical examples of the data before 1958, matched to their own measurements from 1958 onward (where the black line thickens). Figure 2.3 shows the data going back to 1700, before the advent of the modern industrial era.

March 27, 2020

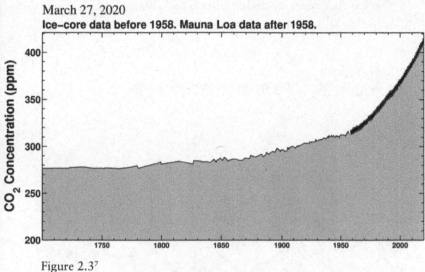

Figure 2.3[7]

Alternatively, one can go back to the dawn of human recorded history, about ten thousand years (Figure 2.4).

March 28, 2020

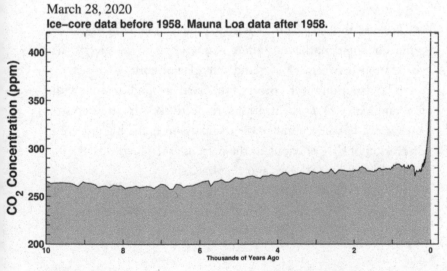

Figure 2.4[8]

The ice core record actually goes back about eight hundred thousand years. Here is the full record (Figure 2.5).

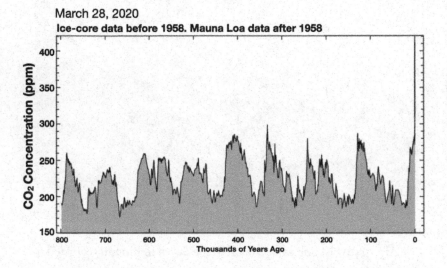

Figure 2.5[9]

Because this last figure spans so much time, it is useful to point out some markers to guide you. Of some relevance is the correlation between ice age and interglacial period, caused by periodic variations in the orbit of the Earth around the sun, with low and high CO_2 concentrations, respectively. Also, an important overall baseline number for comparison is the highest concentration of CO_2 previous to the present era, which was at 300 ppm, about 350,000 years ago (Figure 2.6).

Returning to the original Keeling curve, we can compare it to the global CO_2 emission from human activity, as taken from a figure prepared by the Global Carbon Project (Figure 2.7).

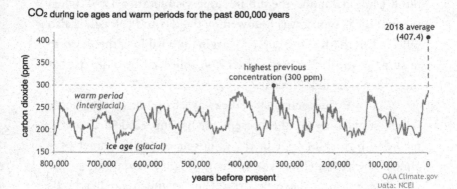

CO₂ during ice ages and warm periods for the past 800,000 years

Figure 2.6[10]

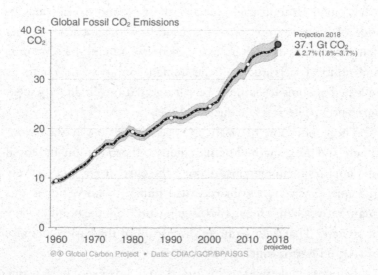

Figure 2.7[11]

The monotonic rise in CO_2 production is compatible with the Keeling data, and the dip in slope visible after 1990, when the first IPCC report was followed by the Kyoto Protocol, is also reflected in the Keeling curve. Marginally visible is the increase in slope in the 1970s preceding 1974, followed by a dip at that time due to the first oil crisis.

What are the immediate takeaways from this data? Some things stand out: (a) the present era is unprecedented in the recorded history of CO_2 in the atmosphere of the planet over almost the past million years; (b) geological variations in the CO_2 concentration have occurred, but at much smaller levels and over much longer timescales than the recently observed rise; (c) those times with higher levels of CO_2 in the atmosphere appear correlated with warming periods, and those times with lower levels with ice ages; (d) the rise began with the beginning of the modern industrial era, and the rate and overall magnitude of increase appears commensurate with global fossil fuel consumption by human industrial activity; and (e) the economic and political vicissitudes of the human condition appear to be reflected at some level in the recent undulations in directly measured atmospheric CO_2 concentrations.

The connection between CO_2 concentration in the atmosphere and the growth of human industrial production and fossil fuel appears unambiguous, making the current era qualitatively and quantitatively new in recorded human history. But is this quantitative change likely to be significant from the point of view of climate? That will require a discussion of the basic dynamics of CO_2 on Earth, which we shall turn to next.

Before completing this historical tour, however, there is one more curve of CO_2 concentration that I find personally compelling. I have shown plots over the recent era, over the industrial era, and over geological eras. The following is a plot over the last two thousand years. All of the modern dramas in human civilization took place over this time, from the rise of Christianity to the fall of Rome, the creation of Islam, the dynasties of China, the medieval era, imperial wars, and in the west, the Enlightenment, the Renaissance, and ultimately the modern technological world, including the atom bomb and two world wars. During this time the human population has grown from perhaps two hundred million to almost eight billion, a fortyfold increase. The energy use per human has increased by a far larger rate. During almost all of these remarkable developments of modern civilization, humans had little global physical impact on the planet—until today.

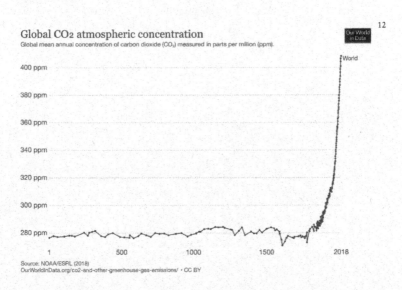

Global CO2 atmospheric concentration
Global mean annual concentration of carbon dioxide (CO₂) measured in parts per million (ppm).

Source: NOAA/ESRL (2018)
OurWorldInData.org/co2-and-other-greenhouse-gas-emissions/ • CC BY

CHAPTER 3

CYCLES AND CYCLES

I celebrate myself, and sing myself,
And what I assume you shall assume,
For every atom belonging to me as good belongs to you.

— WALT WHITMAN, "Song of Myself"

THE HISTORY OF EARTH IS A HISTORY OF OVERLAPPING CYCLES, repeated with great regularity on ever-decreasing timescales. We are relative latecomers to that history, and while we often imagine ourselves as the Masters of the Universe, we are nevertheless the slaves of chemistry.

Every atom that makes up you and me has been recycled throughout the universe and throughout Earth on its cosmic journey. Yes, we are stardust, but we are also intimately connected to everything that has ever lived on Earth and to the very rocks we walk on, water we swim in, and air we breathe.

With the exception of hydrogen, every other element in Earth's atmosphere, and in our bodies, including carbon, oxygen, and nitrogen, was fabricated in the fiery cores of stars. Their explosive deaths in supernova explosions seeded the galaxy for the formation of new stars and the solar systems that formed around them. Hydrogen is the dominant element in the universe, but stars produce significant amounts of carbon, nitrogen, and oxygen, as well as silicon and iron, during their lifetimes. This accounts for the fact that iron and silicon make up much of the interior of rocky planets like Earth, as well as the interior of asteroids and meteoroids that orbit throughout the solar system. Equally important, planets large enough to have atmospheres and hold on to them for billions of years have lighter elements, like carbon, nitrogen, and oxygen, and compounds containing hydrogen, dominating their atmospheres.

From the dawn of terrestrial time, these elements have been recycled throughout the planet. The hellish period of the Earth shortly after the gigantic impact that created the moon and liquified much of the planet is called the Hadean period, after Hades, and for good reason. As the molten rock condensed after the impact within a few thousand years, a dense atmosphere was left behind containing mostly CO_2, hydrogen, and water vapor, with no free oxygen. Estimates are this early atmosphere was almost thirty times denser than our current atmosphere, and the dominant gas was CO_2. This means this initial concentration of CO_2 in the early atmosphere was well over ten thousand times greater than it is today.

CO_2 was reduced to its present abundance as a result of the first great carbon cycle on Earth, the geological cycle. Before the emergence of life, it was the only carbon cycle. The dense CO_2 layer began to decrease as the Earth cooled and liquid water oceans began to cover the Earth's surface. This first bit of chemistry is simple. Carbon dioxide dissolves in water and combines with it to form carbonic acid:

$$H_2O + CO_2 = H_2CO_3$$

The carbonic acid interacts in the ocean, or with the rock, forming substances called carbonates, where a CO_3^{2-} ion combines with things like calcium, magnesium, or iron. These tend to be insoluble in water and sink to the bottom of the ocean floor (like the buildup of scale inside pipes with hard water), effectively removing carbon from the atmosphere-ocean system.

This process could not continue indefinitely if that were the end of the story, as eventually some equilibrium between the carbonates and carbonic acid and dissolved carbon dioxide would occur, including the formation of bicarbonates, which are soluble in water. However, the continental crust of the Earth floats on a moving mantle, causing the continents to drift on large continental plates. At the places where the plates meet, one plate is thrust upward by continental drift and the other plate downward. This downward process is called subduction.

Subduction, beginning at deep ocean trenches, drags material, including the carbonate products of the carbonic acid interactions, down into the Earth. This further drives the removal process of the initial atmospheric CO_2.

If this were the end of the story, the removal of CO_2 from the atmosphere could continue indefinitely. However, as the material is dragged down into the mantle, it is heated, breaking apart the carbonate materials and releasing CO_2 into the mantle. Eventually, after enough CO_2 has been moved from the atmosphere to the mantle, the pressure of CO_2 in the mantle builds up. Excess CO_2 is then released in regions where the crust is thrusting upward due to continental drift. This is where new land rises above the ocean, accompanied by violent volcanos that spew hot rock and gases back into the atmosphere.

Ultimately, over a long enough period, this would create some sort of equilibrium between CO_2 in the atmosphere and CO_2 stored in the Earth. During this process, an average CO_2 molecule in the atmosphere would get recycled through the Earth about once every hundred million years or so.

Over geological time frames (in other words, tens or hundreds of millions of years) this cycle is self-regulating. Roughly speaking, if the CO_2 concentration in the atmosphere increases, more CO_2 dissolves in water, producing more carbonic acid in water. This increases the creation of carbonates that remove CO_2 from the atmosphere/ocean interface, reducing the atmospheric CO_2 ratio. If the CO_2 concentration in the atmosphere falls, it is replenished by the CO_2 released by volcanos based on the previous buildup of carbonates in the crust and mantle.

While it gets me a bit ahead of myself, I can't resist relating here how this process dramatically affected Earth in a fascinating way during a period about six hundred to eight hundred million years ago. Over the past twenty-five years, it has become widely accepted that Earth experienced one or more periods where, due to the merging of continents into a supercontinent and Earth's subsequent ability to reflect sunlight, Earth's surface actually froze over.

I remember when I first heard of this possibility discussed by a young assistant professor at a meeting of the National Academy of Sciences about twenty years ago, I was flabbergasted and skeptical. We think of Earth as a beautiful water world, and even today we classify the possibility of "habitable" planets around other stars by whether they have liquid oceans. And to find out that during at least one relatively recent period in our history, our habitable planet resembled not Mars but the frozen water moons Europa or Enceladus, was stunning. But within a few years of that meeting, the "Snowball Earth" hypothesis, as it became known, was sufficiently well accepted in the community that I included it in the narrative in my book *Atom,* which described a hypothetical cosmic history of a single oxygen atom from the beginning of the universe until its end.

What would a possible Snowball Earth event do to the global geological carbon cycle? With the ocean surfaces frozen, the atmosphere/ocean interface governing the flow of CO_2 from the atmosphere into the ocean would be interrupted. During this period, volcanos would continue to emit deeply stored

CO_2 back into the atmosphere and increase the atmospheric CO_2 concentration.

The way to halt this increasing concentration would be to melt the Earth's surface so CO_2 could once again dissolve into the oceans. Fortunately, the driving mechanism for this melting *was* the very buildup of CO_2 in the atmosphere itself. As we shall discuss in some detail in what follows, CO_2 helps store heat that would otherwise be reflected back into space by Earth. Thus, as the CO_2 buildup during Snowball Earth continued, more heat was stored and less radiated into space, eventually melting the oceans and allowing the geological carbon cycle to once again resume.

The emergence of life on Earth completely changed not only the atmospheric CO_2 concentration but almost everything else, transforming the planet. While likely initially powered by inorganic reactions at the sea floor, eventually life discovered the sun as an energy source. Not only is sunlight ubiquitous at Earth's surface, but more than ten thousand times the energy released from Earth's core to the surface impinges on the same surface each day from the sun.

Photosynthesis uses sunlight to elevate the energy levels of electrons in atoms so those electrons can later do work to power the chemical reactions of life. In the process, as I have already briefly mentioned, it breaks apart water and uses CO_2 as a basic ingredient to build complex hydrocarbons. The waste product of this process is oxygen in the form of O_2 molecules.

The earliest forms of photosynthetic life—cyanobacteria, tiny single-celled objects—provide a literal testament to the idea that even in small groups, working together can change the world. Over time, day after day, year after year, century after century, millennium after millennium, over the first few billion years of the history of this planet, each small puff of O_2 released by every single cyanobacterium contributed to a remarkable conversion of CO_2 to O_2 in the atmosphere. Colonies of cyanobacteria, producing structures called stromatolites—visible as extinct fossils or extant forms in places such as Australia and Mexico—photosynthesized CO_2 and H_2O into O_2. Since O_2 was initially a poison, layers of cyanobacteria would die as these structures built up, and slowly these colonies helped remake the Earth's atmosphere to bring O_2 and CO_2 concentrations close to their present values about a billion years ago.

Why was O_2 a poison? Because it could "oxidize" materials, from iron to organic substances, stealing the free electrons organic materials could use to store energy for later use. It was vitally important that there was little or no free oxygen in the early atmosphere. Only after O_2 levels in the atmosphere rose sufficiently did living systems respond to this new abundance by evolving a mechanism to exploit oxygen rather than succumb to it. Respiration is basically "controlled burning." Living systems release stored oxygen electrons to help produce a substance called ATP—akin to a "battery" universally used by living systems to power their physical processes—during the last phases of the electron-mediated power cycle. As a result, living systems are able to extract over thirty times the energy they would otherwise have been able to generate by photosynthesis. During respiration, incoming O_2 allows organic molecules to be fuels for life, just as during burning it supports the combustion of wood. In both cases oxygen combines with carbon from organic materials to produce CO_2.

Photosynthesis utilizes CO_2, water, and sunlight as fuel sources, while respiration exploits O_2 and returns CO_2 to the atmosphere. These two energy-generation processes of life have recycled CO_2 on Earth in ways that ultimately have governed the dynamics of CO_2, even as its geological cycling through Earth has continued.

As oxygen began its slow but inexorable rise from its near-zero concentration in the atmosphere of the initial, lifeless planet to its current value of about 20 percent of the atmosphere, the CO_2 concentration inexorably fell to its current value of about 0.04 percent of the atmosphere. The carbon that was extracted by photosynthesis was processed into organic materials that built up over the billions of years life has been evolving on the planet. The carbon is now stored in biota on the surface, in the soils of the world, and deep underground.

When first confronted by the claim that human civilization has recently altered the global dynamics of our atmosphere, it seems slightly outrageous. The Earth is vast, and the atmosphere extends tens of miles upward to the borders of space. Every day the sun rises and sets, tides go in, tides go out, to once again paraphrase the ridiculous former Fox broadcaster Bill O'Reilly. We experience the annual cycles as winter turns to spring, spring to summer, summer to fall, and fall again to winter. We experience beautiful sunsets and clear evenings full of stars if we are lucky to be in the right place at the right time. All this reinforces the seeming insignificance of humanity on a global scale.

To be sure, the calm regular rhythm of life is periodically interrupted by natural disasters, but even these reinforce the perception that Mother Nature is far more powerful than we are. With hurricanes, typhoons, snowstorms, floods, droughts, and deluges, the natural world seems too powerful to be dramatically affected by anything everyday human activity can generate. Moreover, if Earth took billions of years to reach its current state, how can we change it in the course of decades?

The key to understanding our human impact here is to realize that we are able to access in a short time, through our technology, quantities of carbon that took literally hundreds of millions of years to build up on Earth before the rise of humans.

First, rather than thinking about the percentage of CO_2 or O_2 in the atmosphere, it is perhaps more useful to think in terms of absolute amounts of each substance. These are traditionally expressed in terms of gigatons (Gt=billions of tons) of material, either as gigatons of CO_2 or gigatons of just carbon. Since the atomic weight of carbon is about 12 atomic mass units (AMU), and oxygen is about 16 AMU, the ratio of the mass of one CO_2 molecule to one carbon atom is about (12 + 2x16=44)/12, or about 3.67. Because different studies quote one or the other, I will sometimes quote carbon and sometimes CO_2 abundances in what follows, and you can use the 3.67 figure to transform from one to the other.

Figure 3.1 is an illustration by Donald DePaolo at Lawrence Berkeley Lab that displays the global carbon cycle during recent history but before the current industrial revolution.

The atmosphere contained about 600 Gt of carbon, or about 2,200 Gt of CO_2. As illustrated here, the carbon in CO_2 gets cycled through Earth's oceans and mantle, and these locations have built up their own carbon reservoirs over time. At this equilibrium level, 600 Gt (the same amount as in the atmosphere, it turns out) resides in the terrestrial biosphere, while about three times that amount, about 1,800 Gt, resides in terrestrial soils. The surface levels of the ocean that are in contact with the atmosphere contain about 1,000 Gt, while the deep ocean, not in contact with the atmosphere, stores a much larger reservoir of 37,000 Gt.

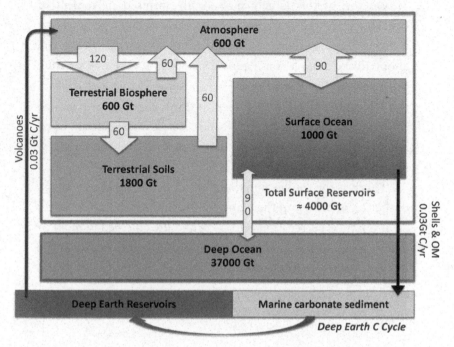

Figure 3.1[13]

During this equilibrium, CO_2 gets recycled in various ways, moving between the atmosphere, the biosphere, soils, and the ocean. As shown in the figure, about 120 Gt/yr. of carbon is taken up by the biosphere, and about 60 Gt/yr. of that is converted into the soils as living organisms die, for example. Similarly, through respiration, decay, fires, and the like, about 60 Gt/yr. is released back into the atmosphere from the biosphere, and an equal amount from soil. About 90 Gt/yr. of carbon is cycled back and forth between the atmosphere and the surface layers of the ocean, and between the surface layers and deep ocean. Altogether, this cycling keeps the carbon abundance in these reservoirs and the atmosphere roughly constant.

On top of this largely biologically inspired movement of carbon, which happens on short timescales, there is the long-term geological carbon cycle. A much smaller amount of carbon, about thirty million tons or so each year, is taken as calcium carbonate and deposited in the ocean floor. As I have described, this is eventually subducted deep into the Earth to be returned to the atmosphere through volcanos. These spew out an equal amount of carbon as CO_2 (less than 1 percent of what humanity now does) back into the atmosphere.

This was the equilibrium that has existed on Earth over much of the recent history of the planet, since atmospheric abundances achieved their current concentration and life emerged in more or less its present form hundreds of millions of years ago. What has happened since, as humans evolved modern technology, is shown in Figure 3.2.

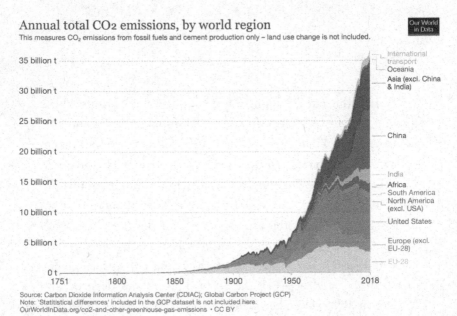

Figure 3.2[14]

At the current time, humanity is emitting over thirty-six billion tons of CO_2 per year into the atmosphere, or in the units used in the previous figure, about ten billion tons of carbon. This graph also breaks down the locations where most of the emissions have taken place. While China has the world's biggest population and is also emitting the most CO_2 at the present time, the USA has been the greatest single emitter when its cumulative contributions over time are considered. I will discuss the impacts of the distribution of CO_2 emissions later but now want to concentrate on the totals.

Up to the present time, human industrial activity over the past sixty years has resulted in the emission of about 400 Gt of carbon into the atmosphere. That is about two-thirds as much CO_2 as previously existed in the atmosphere over the past million years or so. Not all of this has remained in the atmosphere, however, as the Keeling curve demonstrates. The CO_2 abundance in the atmosphere has increased by only about 30 percent since 1960, meaning about 50 percent of the emitted CO_2 has been taken up by a combination of the surface ocean layers and terrestrial soils.

Humanity has efficiently tapped into a much larger reservoir of carbon than has been available at the Earth/ocean-atmosphere interface up to now. By digging into the deep Earth, we have efficiently extracted amounts of carbon that would otherwise have required eons for any previous carbon cycle to generate. To get a sense of this, we can add the human contribution to the global CO_2 balance in an updated block figure (Figure 3.3), along with the net amount of extra CO_2 that gets taken up by ocean and soil.

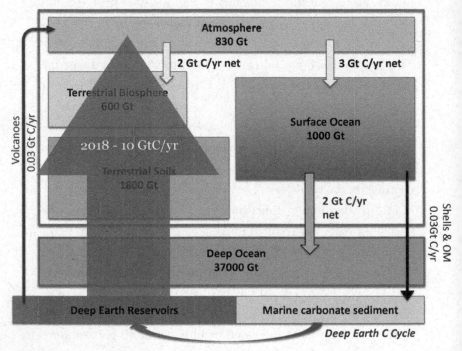

Figure 3.3[15]

Unless we drastically alter the way we generate energy, which today is primarily by burning fossil fuels, during the next 150 years, humanity will emit an additional 1,000 to 4,000 Gt of carbon into the atmosphere, or up to seven times the total amount of CO_2 that existed in the atmosphere before the advent of modern civilization. If this seems staggering, consider an estimate I shall later describe that was made over one hundred years ago: If all of the carbon in the atmosphere in 1900 were compressed into solid form, it would comprise a global layer less than 1 mm thick on the Earth's surface. It does not seem so daunting, therefore, to imagine that humanity's carbon contribution to the atmosphere from industrial activity over the past two hundred years might exceed this amount.

As non-intuitive as it may seem, when it comes to the global terrestrial carbon budget at the planet's surface, the human footprint is anything but negligible.

CHAPTER 4

EARTH'S BLANKET

When I feel the heat, I see the light
— Everett Dirksen

Why isn't the Earth as cold as the space around it?

While not something you may wonder about every day, a version of this question was what the famous mathematician and physicist Joseph Fourier, whose work on what is now called Fourier analysis forms the mathematical basis of much of modern theoretical and experimental physics work today, asked himself in 1824.

More specifically, Fourier reasoned that Earth would radiate much of its heat out into space if something were not trapping that heat—something like a blanket, which keeps us warm at night by stopping us from radiating our own heat out into the room when we are sleeping.

Fourier had been studying the mathematics of heat flow for some time and, in the course of that study, discovered the mathematics that now bears his name. It was in the context of his studies of heat flow that he began to study Earth's temperature. He explored diurnal temperature variations as well as seasonal ones and determined that the planet was warmer than one would expect if the only source of warming came from the sun's radiation, at least if much of that radiation could then be reradiated by Earth out into space.

The Earth needed an insulating blanket, and he reasoned that Earth's blanket of air, which shields us from the harshness of space in many ways, acts like an insulator.

He considered the example of a box with a glass cover. The air inside the box heats up when exposed to sunlight. Anyone who has sat in a car with the windows closed in the summertime has experienced this phenomenon. The specific example that Fourier quoted came from an experiment designed by Horace Bénédict de Saussure, involving a vase with the interior containing black cork and covered by layers of glass separated by air. Thermometers put into the vase and between the layers showed that, when exposed to sunlight, the interior of the vase reached the highest temperature and subsequent layers of air between the glass plates had successively lower temperatures.

His example was a somewhat unfortunate choice, as it became later known as the Greenhouse Effect, coined by Swedish meteorologist Nils Gustaf Ekholm, after greenhouses, which are essentially glass boxes that store heat from the sun to produce a temperate environment for the plants inside. The reason the choice was unfortunate was that greenhouse glass keeps the inside air warm in large part because it traps the warm air heated by the sun inside of it, whereas on Earth warm air is free to rise by convection or transporting heat upward by conduction. Absorbing heat radiation before it escapes is important for both greenhouses and the atmosphere, but that effect is less important in greenhouses than the fact that they trap air.

Fourier recognized this fact but concentrated on the issue of heat radiation when he discussed this experiment. In his (translated) words:

> The theory of this instrument is easy to formulate. It suffices to remark that: (1) the heat acquired is concentrated because it is not dissipated immediately by exchange of air with the surroundings; (2) the heat emanated by the Sun has properties different from

those of dark heat. The rays of this star are for the most part transmitted through the glass without attenuation and reach the bottom of the box. They heat the air and the surfaces which contain it: the heat communicated in this way ceases to be luminous, and takes on the properties of dark radiant heat. In this state, the heat cannot freely traverse the layers of glass which cover the vessel; it accumulates more and more in the cavity enclosed by materials which conduct heat poorly, and the temperature rises to the point at which the incident heat is exactly balanced by the dissipated heat. One could verify this explanation, and render the consequences more evident, if one were to vary the conditions of the experiment by employing colored or darkened glass, and by making the cavities containing the thermometers empty of air. When one examines this effect by quantitative calculations, one finds results which conform entirely to those which the observations have yielded. It is necessary to consider this range of observations and the results of the calculations very carefully if one is to understand the influence of the atmosphere and the waters on the thermometric state of the Earth.

The term "dark heat," as translated here, refers to infrared radiation, discovered by Sir William Herschel in 1800.

In any case, what Fourier described in his antique language is more or less what is understood today to account for the fact that Earth is warmer than it would be without its atmosphere. Actually, two factors come to play: Earth's overall reflectivity and a greenhouse effect in the atmosphere. I will consider each in turn, approaching the subject as a physicist does, by starting with the simplest picture then adding some additional complexity after exposing the key ideas.

Every second of every day, about 1361 watts per square meter crosses every point on a sphere located at Earth's radius, centered on the sun. Since Earth is a sphere and not a disk, not all of the radiation impinges directly on every part of Earth facing the sun. So to estimate the total radiation hitting Earth, we can replace it by the area of a disk πR^2 facing the sun with a radius equal to Earth's radius, R (Figure 4.1).

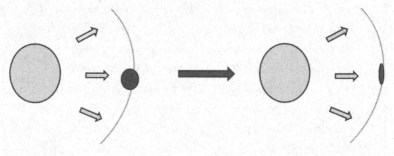

Figure 4.1[16]

Dividing this total by actual Earth's surface area, which is four times as large, the average solar radiation power is therefore reduced by a factor of four to about 340 W/m². In equilibrium, Earth will radiate away into space as much radiation energy as impinges on it from the sun. Otherwise it would continue to heat up. Earth also directly reflects some fraction, a (about 30 percent), of the sun's radiation. So the total rate of energy flowing into Earth from the sun is given by $P_{solar} = 1361 \times (1-a)\pi R^2$. This turns out to be about 10^{17} watts, or about 100,000 terawatts of power. By comparison, humanity uses about 40 terawatts on average. Therefore, more than a thousand times the energy used by all of humanity rains down from the sun each day. If we could just harness even a small fraction efficiently, humanity's energy needs would be more than satisfied.

In any case, to not heat up any further, the Earth has to emit all of this power back into space, which it does by radiating the heat out as a thermal body. It is one of the fundamental laws of physics, as we now understand them, that bodies at some temperature, T, radiate energy out at a rate that goes as the fourth power of temperature, multiplied by a constant that physicists label as the Stefan-Boltzmann constant. I simply label it as k here. The Earth radiates over its entire area, which is given by $4\pi R^2$, so the total power radiated by the Earth at temperature T is given by $P_{Earth} = (4\pi R^2) \times (kT^4)$. To figure out the equilibrium temperature of the Earth, we just equate this quantity to 10^{17} watts and solve for T.

This is a high school-level physics problem, and when we plug in the numbers, we get that $T \approx -18°C$, about $251°K$ (K=Kelvin, or degrees Celsius above absolute zero) or about 0°F. Earth should be frozen solid!

This was the quandary Fourier faced, and the solution he suggested used the fact that a body at this temperature radiates most of its energy in the form of infrared radiation. The wavelength of radiation emitted by hot bodies depends on their temperature. That is why heating elements go from red-hot to blue- or white-hot as their temperature increases. An object with the average temperature of the Earth's surface (about 15°C) radiates in the infrared band, with wavelengths longer than visible light. But, as you can literally see by looking up, the wavelengths of light from the sun are centered in the visible band. Shown in Figure 4.2 are the spectra of light (as predicted by modern physics) from an object at the sun's surface temperature and a "black body" (one that absorbs and re-emits all incoming radiation thermally) at the Earth's surface temperature.

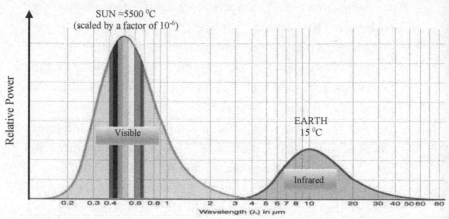

Figure 4.2[17]

The difference in the frequency of the two different types of radiation is important. From the fact that we can see the sun, we know the atmosphere is largely transparent to radiation in the visible band. But it may not be transparent to radiation in the infrared band. In this case it would absorb radiation emitted upward by the surface of the Earth, but not incident radiation coming down from the sun. That would change everything, and this possibility is central to the terrestrial greenhouse effect.

To see this more explicitly, assume that the Earth's atmosphere absorbs and stores some fraction, say b, of that energy. Then instead of radiating a power P_{Earth} into space as before, the Earth would radiate $(1-b) P_{Earth} = (1-b)4\pi R^2 kT^4$. Then, to radiate as much power into space as it receives from the sun,

$$(1-b) P_{Earth} = P_{solar}$$

Since $(1-b) < 1$, this means the power radiated at the surface of the Earth would have to be greater than it would be if b were 0. Since the power radiated is proportional to the fourth power of its temperature, this in turn means the Earth's temperature would have to be greater than it would otherwise be if some of the energy it would otherwise radiate into space didn't get trapped in the atmosphere first.

Absorption of IR radiation is therefore at the heart of the greenhouse effect, although the actual details are more complex. Even though the model I have just presented is simplistic, as physicists are wont to do on a first try, simple models often reveal the key physical factors that are at play. In another book I describe a joke in which a physicist first pictures a cow as a sphere when trying to analyze it. But even that simplistic picture can reveal a lot—about not just cows, but life in general.

We can be a little more realistic by recognizing that the atmosphere itself, if it absorbs heat energy, cannot simply just store it, but it must also radiate energy if it is not to continue heating up. Approximating the entire atmosphere as having a single temperature, it will radiate with a power $P_A = 4\pi R^2\, kT_A^4$. But it will then emit radiation in two directions, both out into space and back down again to Earth (Figure 4.3).

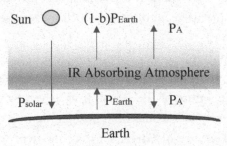

Figure 4.3[18]

For Earth to be in equilibrium with the incoming solar radiation, the sum of the emission from the Earth into space plus the emission from the atmosphere into space must equal the incoming power coming from the sun:

(1-b) $P_{Earth} + P_A = P_{solar}$

For the atmosphere to be in equilibrium with the radiation impinging upon it from the Earth, its total emission, that is, its emission back toward the Earth plus its emission into space, must equal the amount it absorbs. This is a fraction b of the radiation emitted from the Earth.

$bP_{Earth} = 2\,P_A$

Using this relation for P_A in the previous expression yields:

(1-b/2) $P_{Earth} = P_{solar}$

This somewhat more realistic model implies that even if subsequent emission from the atmosphere into space is included in the estimate following absorption of IR radiation from the Earth, the power emitted from the Earth still increases compared to what it would be if the atmosphere absorbed no radiation. The quantitative amount by which it increases is now different, however.

More importantly, this model is only consistent if we assume the temperature of the atmosphere radiating into space is lower than the temperature of the Earth. If it were the same, then the top of the atmosphere would be emitting radiation with the same power as the surface of Earth. This is consistent with $b = 0$, or no absorption of IR radiation by the atmosphere. As we shall see, this requirement is satisfied precisely because the atmosphere is not homogeneous and the top of the atmosphere is both less dense and cooler.

Expecting precise and accurate quantitative estimates on the basis of such a simple model is unrealistic. But using it as a rough guide demonstrates the significant role of absorption in determining the overall magnitude of the observed greenhouse effect on Earth. The average temperature at Earth's surface is observed to be 15°C and not the value of -18°C we derived for an Earth without any absorbing atmosphere. Using this simplistic model, one finds that most of the IR radiation emanating from the Earth's surface would have to be absorbed by the atmosphere—indeed almost 80 percent of it.

Perhaps the main reason detailed quantitative estimates using this model are approximate at best is because the model as I have presented ignores a number of physical effects besides absorption of solar radiation by the Earth's surface and subsequent thermal radiation by the surface and the atmosphere.

First off, while the atmosphere is mostly transparent to the sun's radiation, thankfully it is not fully so, or we might all die of skin cancer from ultraviolet radiation. In total the atmosphere absorbs about 23 percent of the incoming solar radiation, meaning between reflection and atmospheric absorption, only about 49 percent of the incident solar radiation actually makes it down to the surface of the Earth.

Next, the Earth's atmosphere and surface are far from uniform, so temperatures vary greatly from the equator to the poles, and temperatures and absorption rates vary vertically through the atmosphere. Thus, it is a rough approximation to treat the atmosphere as a single homogeneous entity. Also, wind and ocean currents transfer heat laterally around the Earth. In addition, the Earth's thermal response time (in large part due to the huge heat capacity of the oceans) is such that it can take centuries for heat buildup to manifest itself as a rise in temperature, so a radiation imbalance can occur for some time before equilibrium temperatures are reached.

Equally important in changing the overall energy balance is the fact that radiation is not the only way heat is transferred from the Earth to the atmosphere since the Earth's surface is in direct physical contact with the atmosphere. Slightly more than 50 percent of the incoming solar energy that reaches the Earth's surface (or about 25 percent of the total incoming solar flux) leaves the surface through evaporation. The heat goes into evaporating water, which is then stored and can be released later in storms. As I will elaborate on later, this allows heat energy to be transferred from warmer to colder places on Earth and drives an atmospheric "heat engine."

Another 10 percent or so of the solar energy reaching the Earth's surface gets transferred back to the atmosphere by convection. Warm air close to the Earth's surface rises up into the atmosphere, where it releases its energy by expanding and cooling.

This leaves only 35–40 percent of the incoming radiation absorbed at the Earth's surface to be released in the form of thermal radiation. The atmosphere naturally absorbs about ⅓ of this radiation, meaning about 24 percent of the radiation that is absorbed by the Earth's surface ends up being reradiated out into space from the surface. This in turn means only about 12 percent of the total incident radiation from the sun (remember, only 48 percent of that total incident radiation makes it down to the Earth's surface in the first place) gets radiated back into space from the Earth's surface. Since about 30 percent of the total incident radiation is reflected, that leaves almost 60 percent of the incident radiation that must be reradiated into space *by the atmosphere* for the Earth-sun system to remain in thermal equilibrium.

This illustrates a feature that is central to the terrestrial greenhouse effect. Most of the incident solar radiation is absorbed by the Earth's surface, while most of the outgoing radiation is radiated into space at the top of the Earth's atmosphere. This is a consequence of the heat energy stored in the atmosphere, as captured in both of the toy models I described earlier.

Note also that the top of the atmosphere is in thermal radiative equilibrium with about 60 percent of the incoming solar energy flux. As a result, one would expect, using the original estimate we made for the Earth's temperature without an atmosphere (-18°C or 253° Kelvin), that the average temperature in the top layer of the atmosphere where the radiation is emitted would be something like $(0.6)^{1/4}$ x 253°K ≈ -48°C. Happily, this is in accord with observations of the temperature near the top of the troposphere, about 10 km or so above the Earth's surface, below which most of the absorbing part of the atmosphere lies.

With the top of the atmosphere radiating energy upward, at least an equal amount (60 percent of the incident solar flux) must be radiating downward. But the atmosphere actually radiates much more energy down to the surface than it does to outer space. This so-called "back radiation" may seem surprising, as it has been a source of confusion, especially among those who are reluctant to accept the basic physics of the greenhouse effect.

First, note that the atmosphere is warmer at lower altitudes, where it is radiating down to the surface, and one might expect more power emitted from this region. But more importantly, recall that the surface is transferring roughly an additional 35 percent of the solar heat it absorbs up to the atmosphere via evaporation and convection, as well as depositing about 5 percent of its radiant energy output directly to the atmosphere. When you add these up, it turns out that the atmosphere needs to radiate back down to the Earth an additional amount of radiation equivalent to about 40 percent of the solar energy originally incident upon the Earth to not continue heating up. This means it radiates downward an energy equivalent of almost 100 percent of the energy originally incident on the Earth from the sun. The surface, which is far warmer than the atmosphere in general, ends up radiating upward about 116 percent of the solar energy incident on the Earth, which balances the fraction of the original solar flux that makes its way down to the surface plus the energy radiated downward by the atmosphere. Of this upward radiation, about 12 percent or so makes its way back up through the atmosphere and outward directly to space, and the remaining 104 percent gets absorbed by the atmosphere.

If the percentages seem confusing, here is a visualization of the Earth-atmosphere energy balance when the two are in equilibrium with incoming solar radiation, provided by the National Oceanic and Atmospheric Administration (NOAA) in Figure 4.4.

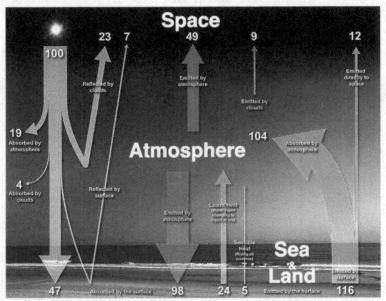

Figure 4.4[19]

It is worth stressing that we can actually measure all of these quantities, so you don't have to take them on faith, even if the general features are already predicted by the simplest models.

Downward-facing satellites can measure the upward-going radiated power emanating from the atmosphere and the Earth's surface. Also, because the spectrum of light incoming from the sun is different than the thermal spectrum of radiation emitted by the atmosphere, upward-facing detectors at the Earth's surface can distinguish between them. Here, for example, are the data from three days of "downward longwave radiation" (the colder atmospheric—or long wavelength—component of downward radiation) measurement in one mid-latitude location in the US, with the power in W/m^2 shown in Figure 4.5.

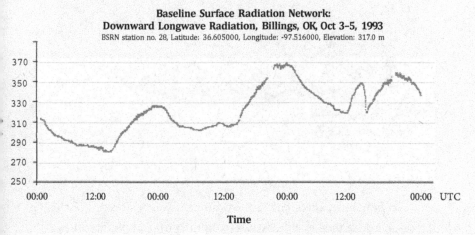

Baseline Surface Radiation Network:
Downward Longwave Radiation, Billings, OK, Oct 3–5, 1993
BSRN station no. 28, Latitude: 36.605000, Longitude: -97.516000, Elevation: 317.0 m

Figure 4.5[20]

While this component varies just as the incident solar flux does, with the highest magnitude at low latitudes and the lowest magnitude at high latitudes, at mid-latitudes it achieves a value close to its average value, which one can see here is indeed comparable in power to the total average incoming solar power impinging on the Earth, as predicted.

Also, for reference, here is satellite data from Sept. 2019 showing the net radiation (incoming sunlight minus reflected light and outgoing heat) flowing into the Earth (Figure 4.6). Note again that near the equator more radiation is absorbed than emitted or reflected, and the opposite is true near the poles. Mid-latitudes are roughly in balance.

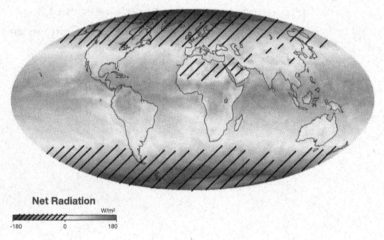

Net Radiation

Figure 4.6[21]

To conclude, then, just so we don't lose the forest for the trees, here are the key takeaways from the physical picture I have presented thus far.

1. The heart of the greenhouse effect, first proposed by Fourier and confirmed by the fact that the surface temperature of Earth is warmer than it would otherwise be, is as follows: If the atmosphere absorbs heat energy

that would otherwise be radiating out into space by Earth, the resulting surface temperature of Earth will be warmer than the atmosphere. As a consequence, it will be warmer than it would otherwise be in the absence of an atmosphere. The atmosphere indeed acts like a blanket.

2. Observations of the temperature of the surface and the atmosphere and measurements of radiant energy flows upward and downward are consistent with this fundamental understanding of the greenhouse effect. This understanding derives from longstanding and well-understood physics arguments. Basic model predictions and observational data combine to establish the basis of this physical picture of the greenhouse effect.

3. While a precise determination of how effective a blanket the atmosphere will provide requires detailed modeling of the atmosphere itself, the generic features and consistent quantitative estimates of the greenhouse effect derive from the following basic principles:

   a. In thermal equilibrium: energy in=energy out.

   b. Thermal bodies radiate with power that is proportional to the fourth power of their temperature.

This basic picture is not controversial. What comes next depends on the details....

CHAPTER 5
THE BIRTH OF CLIMATE CHANGE

*Persistence is the character of man
as carbon is to steel.*

NAPOLEON HILL

WHILE FOURIER REALIZED THAT THE ABSORPTION OF OUT-
ward-going infrared heat energy is the key to what became
known as the greenhouse effect, he did not propose what the
source of this absorption was. Nor did he provide quantitative
estimates of what it might be.

The quantitative science underlying the greenhouse effect
began with Irish physicist John Tyndall, who was born just
around the time Fourier was beginning to ponder the question
of what governed the Earth's temperature. In the late 1840s he
and a colleague, who both taught at an English boarding school,
had the good sense to move to Germany to study. German uni-
versities were far ahead of Britain in experimental science at the
time. By the time Tyndall returned in 1851, he was well prepared to begin a career in experimental chemistry and physics,
where his early work focused on magnetism, one of the subjects
of greatest interest at the time. In 1853 he obtained a professor-
ship at the Royal Institution, led by Michael Faraday, one of the
greatest experimental physicists of the nineteenth century, whose
work established the modern theory of electromagnetism. A
decade later, upon Faraday's retirement, Tyndall was appointed
his successor.

A lifelong interest in mountain climbing and glaciology led Tyndall to explore the heating effects of sunlight and study the earlier work of Fourier and others proposing that the absorption of infrared light by the atmosphere might ultimately heat the Earth's surface. Ever the experimentalist, in 1859 Tyndall began a decade-long set of experiments on the absorption of thermal radiation by different gases.

He began experiments in May of that year with a setup that consisted of a known heat source, a tube containing the gas he wanted to study, and at the end of that an object called a thermopile, which converts thermal energy into an electrical signal he could measure. By measuring the thermal energy emanating from the tube, he could determine how much energy was absorbed by the gas.

Incidentally, this was an early application of what is called absorption spectroscopy, which, turned on its head, was ironically essentially the same tool used by Keeling to first accurately measure the fraction of CO_2 in the atmosphere. Tyndall helped first determine the absorption properties of CO_2, and his work was followed by a century of more refined studies. When Keeling did his work, he could compare the absorption of infrared radiation by atmospheric samples to the known absorption properties of test samples with predetermined CO_2 fractions to determine the CO_2 fraction in the atmosphere. A hundred years earlier, Tyndall actually used his own method to invent a system for measuring the CO_2 fraction in human breath, a method that is still in use today.

Returning to Tyndall, by the end of May, he enthusiastically reported his results to the Royal Society:

With the exception of the celebrated memoir of M. Pouillet on Solar Radiation through the atmosphere, nothing, so far as I am aware, has been published on the transmission of radiant heat through gaseous bodies. We know nothing of the effect even of air upon heat radiated from terrestrial sources.

By June 10 he had firmly established the experimental physics of the greenhouse effect. As he put it in a Royal Society lecture, regarding solar heat:

> when the heat is absorbed by the planet, it is so changed in quality that the rays emanating from the planet cannot get with the same freedom back into space. Thus the atmosphere admits of the entrance of solar heat; but checks its exit, and the result is a tendency to accumulate heat at the surface of the planet.

Tyndall studied the infrared absorption by nitrogen, oxygen, water vapor, carbon dioxide, ozone, and methane, among other gases in the atmosphere, and was the first to get quantitative data on this absorption. His chief result was that the main gases, oxygen and nitrogen, did not have any significant absorptions, but that water vapor is the strongest absorber of infrared radiation in the atmosphere. He also demonstrated that the gases that absorb infrared radiation are also infrared emitters, establishing, as we have seen, another key component of the greenhouse effect.

There is a somewhat simple reason why O_2 and N_2 molecules don't absorb much infrared radiation, while CO_2 and H_2O do. Absorption of infrared light occurs when the incident radiation excites vibrations of molecules. It can excite only those molecules whose vibrations also emit infrared radiation.

Molecules are bound when atoms share electrons. If there are different atoms in the molecule, the shared electrons can be attracted more closely to some atoms, making them slightly negative, and the other partner atoms slightly positive, producing what are called electric dipoles. O_2 and N_2 molecules contain two identical atoms and so no atom is more negatively or positively charged than the other atom, and when they vibrate, they do not generate electromagnetic radiation. In CO_2 and H_2O, shared electrons bind more closely to the oxygen atoms, making them slightly negative. As these molecules vibrate, the movement of charges can produce infrared radiation, and thus infrared radiation can excite vibrations in these molecules. In CO_2 there are two modes of vibration that result in the production or absorption of infrared radiation, an asymmetric stretching vibration, and a bending vibration, shown in Figure 5.1:

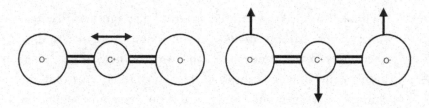

Figure 5.1[22]

This bending vibration is slower, so lower-frequency infrared radiation can excite these vibrations. As such, CO_2 and H_2O preferentially absorb in the infrared.

With Tyndall's results, by the 1860s the emerging experimental basis of the greenhouse effect was largely demonstrated. The astute reader will notice, however, that CO_2 had not yet played a prominent role. Rather, Tyndall demonstrated that it was water vapor that dominates the warming of the Earth, although small amounts of hydrocarbons and CO_2 could have a significant effect. (Incidentally, water vapor was initially responsible for the hellish heat on Venus and will be responsible for a runaway greenhouse effect on Earth in about two billion years or so, by which time the solar luminosity will have increased by about 15 percent. The increased temperatures this will produce will cause the world's oceans to begin to evaporate, putting more infrared-absorbing water vapor in the atmosphere, which will further heat up the Earth, causing more evaporation, and so on. After that time, the surface temperature of the Earth will resemble that on Venus now, in excess of 450°C, or hot enough to melt lead.)

Enter Svante August Arrhenius, a Swedish physical chemist and Sweden's first Nobel laureate in 1903. The work that earned him the prize involved demonstrating that when salts dissolve in water, they dissociate into separate charged particles, which Faraday had called ions. Later on, he demonstrated that for most chemical reactions to proceed, additional heat energy is required to overcome an energy barrier that otherwise stops separate molecules from reacting and developed an equation that still bears his name for a relationship between this "activation" energy and reaction rates. Still later he did seminal work in the early development of biochemistry and even proposed the theory now known as *panspermia*, which suggests that life first arrived on Earth through the transport of spores from earlier life-harboring planets.

So, he was no slouch. But he seems to have irritated his colleagues from his student days onward. His PhD thesis, which contained the germ of the idea for which he would win the Nobel Prize, was rated as "third-class" by his professors, and even that was done only after his defense that the original grade of "fourth-class" should be raised. He became a professor of physics at Stockholm University, but only after overcoming the opposition of his colleagues, and was elected to the Swedish

Academy of Science in 1901, but only after further opposition from his colleagues.

He weathered the opposition well, however. He became rector of Stockholm University in 1896 and later on rector of the Nobel Institute itself. I find it intriguing that he actually helped set up the Institute and the Nobel Prizes, and was a member of the Nobel Committees on Physics and Chemistry, then two years later was the first Swede to be awarded the Prize. Happily, even if there may have been a bit of nepotism involved, his contributions to the fields of chemistry and physics were nevertheless important and lasting.

In 1896 Arrhenius turned his considerable skill in physical chemistry to the question of climate change. He knew of the work of Fourier and Tyndall (and also Claude Pouillet) demonstrating that a greenhouse effect could warm the Earth. The physical question that drove him appeared to be his interest in the appearance and disappearance of ice ages. As he tellingly put it in his groundbreaking 1896 paper, "On the Influence of Carbonic Acid in the Air upon the Temperature of the Ground," published in the *Philosophical Magazine and Journal of Science*,

> I should certainly not have undertaken these tedious calculations if an extraordinary interest had not been connected with them. In the Physical Society of Stockholm there have been occasionally very lively discussions on the probable causes of the Ice Age; and these discussions have, in my opinion, led to the conclusion that there exists as yet no satisfactory hypothesis that could explain how the climatic conditions for an ice age could be realized in so short a time as that which has elapsed from the days of the glacial period.

He therefore became interested in exploring how possible periodic variations in the atmosphere could change the greenhouse effect and therefore also the Earth's temperature over "short" time intervals. Naturally he was concerned with whether variations large enough to affect the climate might be possible due to natural processes on the Earth's surface. Here, he translated into English the lesser-known work of his colleague Arvid Högbom, who estimated how the CO_2 (referred to as carbonic acid at the time in deference to the substance formed when CO_2 dissolves in water) concentration then in the atmosphere might compare to the carbon stored in living systems, and I quoted his result earlier in the book. I find Högbom's estimate to be a wonderful example of the type of "order of magnitude" arguments that are so important and powerful in physics. They allow us to quickly discern what are the crucial physical factors we should focus on and which are likely to be quantitatively irrelevant—namely the type of thinking that motivates this book. For that reason, I present his discussion verbatim here:

> On the supposition that the mean quantity of carbonic acid in the air reaches 0.03 vol per cent. [Note by me: this agrees extremely well with the current estimate that CO_2 comprised .03 percent of the atmosphere in 1900 as can be seen by the graphs in chapter 2], this number represents 0.045 per cent. by weight, or 0.342 millim. partial pressure, or 0.466 grams of carbonic acid for every cm^2 of the Earth's surface. Reduced to carbon this quantity would give a layer about 1 millim. thickness over the Earth's surface. The quantity of carbon that is fixed in the living organic world can certainly not be estimated with the same degree of exactness; but it is evident that the

numbers that might express this quantity ought to be of the same order of magnitude, so that the carbon in the air can neither be conceived of as very great nor as very little, in comparison with the quantity of carbon occurring in organisms. With regard to the great rapidity with which the transformation in organic nature proceeds, this disposable quantity of carbonic acid is not so excessive that changes caused by climatological or other reasons in the velocity and value of that transformation might not be able to cause displacements of the equilibrium.

It may be over one hundred years old, but this still stands as a beautifully simple and powerful order-of-magnitude argument. It demonstrates that global atmospheric CO_2 might be affected, in principle, at significant levels by living systems on biological, and not geological, timescales. I was pleased to learn of it while researching for this book.

The first thing Arrhenius did was ingeniously use infrared observations of the moon taken by Samuel Pierpont Langley to calculate how much infrared radiation is absorbed by CO_2 and water vapor in the atmosphere. He focused on CO_2 as well as water vapor in part because Tyndall had shown that while water vapor was the dominant greenhouse gas, CO_2 and some other trace gases could also have some impact. Also, researchers who followed after Tyndall argued that CO_2 played a more significant role than Tyndall had accorded it. Arrhenius came up with the absorption table (Figure 5.3), which I also reproduce here because the amount of effort he devoted to developing it was considerable.

TABLE III.—*The Transparency of a given Atmosphere for Heat from a body of* 15° C.

$\frac{\rightarrow H_2O}{\downarrow CO_2}$	0·3.	0·5.	1·0.	1·5.	2·0.	3·0.	4·0.	6·0.	10·0.
1	37·2	35·0	30·7	26·9	23·9	19·3	16·0	10·7	8·9
1·2	34·7	32·7	28·6	25·1	22·2	17·8	14·7	9·7	8·0
1·5	31·5	29·6	25·9	22·6	19·9	15·9	13·0	8·4	6·9
2	27·0	25·3	21·9	19·1	16·7	13·1	10·5	6·6	5·3
2·5	23·5	22·0	19·0	16·6	14·4	11·0	8·7	5·3	4·2
3	20·1	18·8	16·3	14·2	12·3	9·3	7·4	4·2	3·3
4	15·8	14·7	12·7	10·8	9·3	7·1	5·6	3·1	2·0
6	10·9	10·2	8·7	7·3	6·3	4·8	3·7	1·9	0·93
10	6·6	6·1	5·2	4·3	3·5	2·4	1·8	1·0	0·26
20	2·9	2·5	2·2	1·8	1·5	1·0	0·75	0·39	0·07
40	0·88	0·81	0·67	0·56	0·46	0·32	0·24	0·12	0·02

Figure 5.2[23]

His units took me some time to figure out. For CO_2, the units are relative: "1" means the amount of CO_2 that would be encountered by a light ray then traversing vertically through the atmosphere, where it passes through the smallest mass of air. The units for water vapor are a bit more standard: "1" represents the amount of water vapor traversed by a vertical ray of light for a humidity of 10 grams/cubic meter at the Earth's surface, which he argued was the mean humidity of the air. Note that Arrhenius therefore estimated that about 70 percent of infrared radiation traveling vertically upward was absorbed by the atmosphere at that time.

He used this data and combined that with radiation equilibrium arguments similar to those I described in the last chapter and calculated how the expected equilibrium surface temperature might be expected to vary as one changed the CO_2 concentration. In several ways, his models were somewhat more complex than I previously described.

First, Arrhenius didn't assume a uniform temperature in the atmosphere and explicitly demonstrated a fact that I alluded to before, namely that a colder upper atmosphere and warmer lower atmosphere will increase the surface temperature because less radiation will be emitted from the top of the atmosphere and heat will be more effectively concentrated near the ground. He also attempted to incorporate the effects of clouds on temperature by recourse to data.

Finally, as far as I can tell, Arrhenius explicitly accounted for the increase in absolute humidity (and thus, the increase in absorption by water vapor) that occurs for latitudes with higher average temperatures when determining how a variation in the CO_2 abundance would change surface temperatures by using measured humidity and temperature at various locations in his calculations. He didn't include, however, an effect that can amplify the impact of rising CO_2 on temperature. Namely, as rising CO_2 concentrations produce rising temperatures, there can be more evaporation of water, increasing the vapor content of the air, and thus further increasing infrared absorption by the atmosphere. But it was pretty impressive for a first pass. Figure 5.4 shows his results, which, though overestimates, nevertheless hold up roughly under modern scrutiny.

As you can see from the table, for an increase of 50 percent in the CO_2 concentration in the atmosphere, he estimated a relatively uniform increase of about 3–3.5°C, whereas for a doubling of CO_2 he predicted an increase of 5–6°C. In all cases he predicted the temperature increase would be larger at the poles than at the equator.

Being the consummate physicist, however, Arrhenius wasn't just satisfied with a table. He preferred a more general rule, or law, he could derive. He did that by interpolating the results in his table. It is also worthwhile to note that the last three columns

TABLE VII.— *Variation of Temperature caused by a given Variation of Carbonic Acid.*

Latitude.	Carbonic Acid=0·67.					Carbonic Acid=1·5.					Carbonic Acid=2·0.					Carbonic Acid=2·5.					Carbonic Acid=3·0.				
	Dec.-Feb.	March-May.	June-Aug.	Sept.-Nov.	Mean of the year.	Dec.-Feb.	March-May.	June-Aug.	Sept.-Nov.	Mean of the year.	Dec.-Feb.	March-May.	June-Aug.	Sept.-Nov.	Mean of the year.	Dec.-Feb.	March-May.	June-Aug.	Sept.-Nov.	Mean of the year.	Dec.-Feb.	March-May.	June-Aug.	Sept.-Nov.	Mean of the year.
70	-2.9	-3.0	-3.4	-3.1	-3.1	3.3	3.4	3.8	3.6	3.52	6.0	6.1	6.0	6.1	6.05	7.9	8.0	7.9	8.0	7.96	9.1	9.3	9.4	9.4	9.3
60	-3.0	-3.2	-3.4	-3.3	-3.22	3.4	3.7	3.6	3.8	3.62	6.1	6.1	5.8	6.1	6.02	8.0	8.0	7.6	7.9	7.87	9.3	9.5	8.9	9.5	9.3
50	-3.2	-3.3	-3.3	-3.4	-3.3	3.7	3.8	3.4	3.7	3.65	6.1	6.1	5.5	6.0	5.92	8.0	7.9	7.0	7.9	7.7	9.5	9.4	8.6	9.2	9.17
40	-3.4	-3.4	-3.2	-3.3	-3.32	3.7	3.6	3.3	3.5	3.52	6.0	5.8	5.4	5.6	5.7	7.9	7.6	6.9	7.3	7.42	9.3	9.0	8.2	8.8	8.82
30	-3.3	-3.2	-3.1	-3.1	-3.17	3.5	3.3	3.2	3.5	3.47	5.6	5.4	5.0	5.2	5.3	7.2	7.0	6.6	6.7	6.87	8.7	8.3	7.5	7.9	8.1
20	-3.1	-3.1	-3.0	-3.1	-3.07	3.5	3.2	3.1	3.2	3.25	5.2	5.0	4.9	5.0	5.02	6.7	6.6	6.3	6.6	6.52	7.9	7.5	7.2	7.5	7.52
10	-3.1	-3.0	-3.0	-3.0	-3.02	3.2	3.2	3.1	3.1	3.15	5.0	5.0	4.9	4.9	4.95	6.6	6.4	6.3	6.4	6.42	7.4	7.3	7.2	7.3	7.3
0	-3.0	-3.0	-3.1	-3.0	-3.02	3.1	3.1	3.2	3.2	3.15	4.9	4.9	5.0	5.0	4.95	6.4	6.4	6.6	6.6	6.5	7.3	7.3	7.4	7.4	7.35
-10	-3.1	-3.1	-3.2	-3.1	-3.12	3.2	3.2	3.2	3.2	3.2	5.0	5.0	5.2	5.1	5.07	6.6	6.6	6.7	6.7	6.6	7.4	7.5	8.0	7.6	7.62
-20	-3.1	-3.2	-3.3	-3.2	-3.2	3.2	3.2	3.4	3.3	3.27	5.2	5.3	5.5	5.4	5.35	6.7	6.8	7.0	7.0	6.87	7.9	8.1	8.6	8.3	8.22
-30	-3.3	-3.3	-3.4	-3.4	-3.35	3.4	3.5	3.7	3.5	3.52	5.5	5.6	5.8	5.6	5.62	7.0	7.2	7.7	7.4	7.32	8.6	8.7	9.1	8.8	8.8
-40	-3.4	-3.4	-3.3	-3.4	-3.37	3.6	3.7	3.8	3.7	3.7	5.8	6.0	6.0	6.0	5.95	7.7	7.9	7.9	7.9	7.82	9.1	9.2	9.4	9.3	9.25
-50	-3.2	-3.3	—	—	—	3.8	3.7	—	—	—	6.0	6.1	—	—	—	7.9	8.0	—	—	—	9.4	9.5	—	—	—

Figure 5.3[24]

for highest values of CO_2 were actually obtained by extrapolating the original data of Langley beyond the domain in which the data were taken.

That said, his result was clear from the table. *A regular change in temperature of about 3°C would be expected each time the value of the concentration of CO_2 increased by a factor of 3/2.* As Arrhenius stated, "if the quantity of carbonic acid increases in geometric progression, the augmentation of temperature will increase nearly in arithmetic progression." This relationship, where one quantity increases linearly while the other quantity increases as a power law (that is, by 3/2 between 0.67 and 1, by $(3/2)^2$ between 0.67 and 1.5, and by $(3/2)^3$ between 0.67 and 2.25), can be mathematically shown to imply that the first quantity (such as surface temperature) varies as the logarithm of the second quantity (CO_2 concentration).

Arrhenius not only pushed the limits of the validity of Langley's data in his extrapolation, he ignored Langley's warning that his data should be taken with a grain of salt. Nevertheless, after a previous century of scientific effort devoted to figuring out how to merely understand the average temperature at the Earth's surface, Arrhenius's goal of exploring how possible changes in that temperature might be induced by a change in atmospheric content was particularly novel and ambitious. And to be fair, he used what was then the best data and theory available to him. As we shall see, the basic framework of his results still stands today. Not bad for work done in 1896.

Interestingly, while Arrhenius is usually credited for being the first person to conclude that CO_2 emissions by human industrial activity were potentially large enough to heat up the planet, his fundamental work was neither motivated by this, nor did he make this claim in his landmark scientific paper. As he stated there, he was motivated by an interest in understanding both the origin of ice ages and how the ice ages ended, and suspected that varying CO_2 concentration might explain these variations in the Earth's temperature. Ultimately, however, he was not able to suggest a precise mechanism for how CO_2 might vary by the requisite amounts over geological time.

Nevertheless, he was aware, from his colleague Arvid Högbom, that the consumption of coal at the time—about five hundred million tons of coal per year—would emit an amount equal to about 1/1000th of the CO_2 then present in the atmosphere. Högbom argued that this additional contribution was easily compensated by the formation of carbonate compounds by weathering processes, though as I described earlier, we now know this latter process is balanced geologically by CO_2 emissions of volcanos. Interestingly, if you take Högbom's estimate and assume roughly a century of coal burning at the rate he suggested, this would increase the atmospheric CO_2 concentration by about 10 percent, were it to all remain in the atmosphere. This is more or less the increase that did occur between 1750 and 1900 (from 275 to 300 ppm) as seen in the extended Keeling curves I displayed earlier.

In a public lecture Arrhenius gave a few years after his scientific paper was presented, he didn't hesitate to extrapolate Högbom's estimate into the future. When he accounted for absorption of CO_2 by the oceans, he estimated that, at the then-current rate of fossil fuel burning, the CO_2 concentration in the atmosphere would double in three thousand years, and global temperatures would increase by 3–4°C. He extolled this possibility as a boon for humanity. Not only would we avoid another ice age, but it would "allow our descendants, even if they only be those of a distant future, to live under a warmer sky in a less harsh environment than we were granted." As a resident of Scandinavia, global warming seemed to him a most welcome prospect!

Arrhenius expanded on these ideas in two books written over the next decade, earning him a reputation today as the first scientist to predict human-induced global warming. Poor Mr. Högbom, whose estimates provided the basis of Arrhenius's claims, has been largely buried in the dustbin of history.

Credit aside, Arrhenius was the first to construct an empirically derived model that allowed for the possibility of understanding how sensitive our climate may be to changes in the abundance of a molecule whose concentration in the atmosphere was less than one part in a thousand, something that without sufficient intellectual preparation, may seem absurd. For that, and the fact that the basic features of his model and the overall magnitude of his predictions remain relevant today, he certainly deserves to be remembered.

CHAPTER 6

FORCING THE ISSUE

I was not born to be forced.
I will breathe after my own fashion.
Let us see who is the strongest.

—HENRY DAVID THOREAU

SO FAR, WE HAVE ESTABLISHED:

a) CO_2 levels have been measured to be increasing yearly at a level consistent with emissions produced by human industrial activity. Current levels are unprecedented over the past one million years or so and more than 30 percent higher than at any other period during that time.

b) CO_2 is measured to be a greenhouse gas, in that it absorbs infrared radiation emitted by the Earth's surface. Along with water vapor, it is quantitatively responsible for keeping the Earth's surface temperate over its history, right up to the present.

c) Increasing the absorption in the atmosphere of outgoing radiation from the Earth's surface increases the surface temperature of the Earth. This is the original greenhouse effect, misnamed perhaps, but nevertheless the physics of which is well established both theoretically and experimentally.

These facts led to the conclusion that the increasing CO_2 concentration in our atmosphere should lead to further warming of the Earth's surface. The key question is how much? Was Svante

Arrhenius correct in estimating 3°C of warming if the CO_2 concentration hits 450 ppm, which, at current rates, will occur by 2050? And even if this were true, what impact would a 3°C rise have on the globe?

These two questions have driven much of the internationally coordinated climate research program that has been carried out over the past sixty years. They have also led to a furor of debate among the public and governments about how to respond to this research since Keeling made his first measurements in 1960.

They are not easy questions to give precise answers to. The general principles are clear and unambiguous, but like much of science, when you want to go beyond a rough understanding to get things right at the percent level, the devil is in the details.

Even in 1900, after Arrhenius first presented his predictions, there was almost an immediately significant academic backlash. The most forceful response came from Knut Ångström, a physics colleague at Uppsala University. Ångström's name carried weight, both due to his genealogy and his scientific expertise. His father was a famous Swedish physicist, after whom the unit of atomic spacing, $(10^{-10}$ m$)$ is named (I confess that I thought this unit was named after Knut until I looked it up recently). More important, his area of expertise was in the measurement of solar radiation.

Ångström based his objections both on theoretical and experimental grounds. He argued that the absorption of infrared radiation was already so great—'saturated' is the technical term—that increasing the abundance of any greenhouse gas, be it water or CO_2, would make little difference. After all, if you are designing a sunshade, and a sheet of wood a millimeter thick already blocks all the sunlight, you don't gain anything by machining the wood any thicker.

Ångström's specific arguments were twofold. First, he had observed the frequency range of absorption by water vapor, and when he measured the same thing for CO_2, he found that the absorption frequency bands appeared to essentially overlap. Since water vapor is far more abundant in the atmosphere, and is also a stronger greenhouse gas, it essentially completely blocks certain wavelengths of IR radiation in the atmosphere. What then is to be gained by increasing the CO_2 abundance?

Second, he added experimental insult to theoretical injury. In 1900 his laboratory assistant, John Koch, performed an experiment with a tube of pure CO_2 gas 30 cm long. This, he claimed, would have contained the same amount of CO_2 as a column of air from the ground up to the top of the atmosphere (remember that even now CO_2 is only 410 parts per million of the total atmosphere). Koch found that when he reduced the amount of CO_2 in the tube by a factor of three, there was no perceptible change in the percentage of IR radiation absorbed by the tube.

These results both turned out to be wrong yet nevertheless carried great weight with the scientific community at the time. Almost fifty years passed before Arrhenius's ideas resurfaced and were vindicated.

Let's deal with the H_2O/CO_2 question first. Precise measurements of the spectrum of infrared absorption by H_2O, which were simply not possible at the time Ångström made measurements, show that what were interpreted at the time as wide bands of H_2O absorption overlapping with CO_2 are in fact a set of narrow spikes separated by valleys. In the following illustration of the absorption spectra (Figure 6.1), you can see that at wavelengths greater than ten micrometers, the CO_2 absorption overlaps with water vapor absorption, and the CO_2 absorption essentially fills in the gaps between the H_2O peaks:

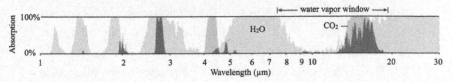

Figure 6.1[25]

This effect gets even more pronounced at high altitudes, where, as I will remind you, the radiation of heat energy from the atmosphere into space occurs. At these altitudes, where the density of air is thin, there are fewer collisions between molecules. Collisions broaden the absorption lines of compounds, so at high altitude the peaks become narrower and the valleys deeper. All of this implies that even in more humid air, where IR absorption in various frequency ranges by H_2O may be saturated, there is still plenty of frequency real estate where CO_2 enhances absorption. And at high altitudes not only do the H_2O absorption lines become narrower, the H_2O abundance itself also falls off to almost zero, whereas CO_2 is well mixed throughout the atmosphere, meaning its absorption becomes even more important at these altitudes.

Now, what about Koch's experiment, and in particular the claim that once CO_2 absorption is saturated, changing the amount of CO_2 won't further increase the absorption? It was incorrect on two counts, but in any case it turns out to be largely irrelevant to the process that drives what has become known as "radiative forcing"—the central parameter that today enters into estimates of global climate change.

The experimental problems with Koch's experiment involved the source of radiation and the length of the tube used. Koch used as his source of radiation essentially boiling water at a temperature of 100°C, well above that of either the Earth or the atmosphere, so the spectrum of the radiation being absorbed would be different. He also used a single 30 cm long tube, and the way he searched for different absorption as a function of CO_2 abundance was to change the CO_2 pressure in the tube. As it happens, the 30 cm tube also underestimated the actual amount of CO_2 that radiation would traverse in the atmosphere—by a factor of five. Countering this is the fact that using CO_2 at ground level atmospheric pressure broadens the absorption lines of CO_2 compared to absorption rates through much of our much lower-pressure atmosphere. Also, using pure CO_2 further broadens the CO_2 absorption lines compared to what would occur in an atmosphere in which trace amounts of CO_2 are mixed with N_2 or O_2. Putting these two competing factors together implies that the appropriate length of tube would have been about one meter to recreate the absorption in our atmosphere.

But neither of these factors would have significantly affected what should have been the correct interpretation of the experiment. The faulty conclusion lied less in execution than in the two assumptions that went into the experimental design.

What Ångström didn't take into account in his notion of saturation was the actual frequency dependence of CO_2 absorption, and the fact that absorption is not an all-or-nothing process. At any frequency, the fraction of transmitted light falls exponentially with the number of CO_2 molecules encountered on the trajectory through the atmosphere. Effectively, saturation occurs when some large fraction of the incoming light is absorbed, say 90 percent. This value can be reached either by a large absorption coefficient of CO_2 at the frequency in question or by increasing the number of CO_2 molecules encountered on the trajectory of light.

Next, the frequency dependence of the CO_2 absorption affects saturation. Adapting an argument from a former *Bulletin of the Atomic Scientists* colleague, Oxford physicist Raymond Pierrehumbert, we can approximate a broad peak in absorption spectrum for CO_2 in some frequency band as looking like Figure 6.2.

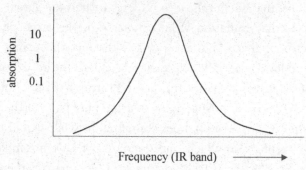

Figure 6.2[26]

Say we consider absorption for all frequencies for which the absorption coefficient is greater than two as being sufficiently saturated so almost all the radiation is absorbed (Figure 6.3).

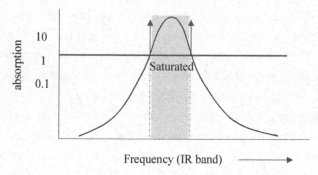

Figure 6.3[27]

Now, if we increase the CO_2 concentration by a factor of, say, four, then all frequencies for which the absorption coefficient is now greater than a value four times smaller, or ½, will be saturated (Figure 6.4).

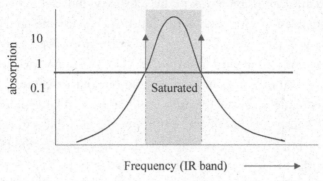

Figure 6.4[28]

What increasing the CO_2 concentration does, therefore, is increase the *range* of frequencies that can have saturated absorption, and this increases the total rate of IR absorption by CO_2 in the atmosphere. So Ångström's concern about saturation was a red herring. Absorption can increase with increasing CO_2 concentration *even if absorption at some frequencies is already saturated*.

What is the effect of this increase? For regions of the atmosphere where the CO_2 density might be high enough, so that effectively CO_2 absorption would be saturated at all frequencies, there wouldn't be any effect. (In fact, modern measurements have since shown there aren't such regions.) However, as one moves up higher in the atmosphere, the atmosphere gets less dense, and with that density reduction, the number of CO_2 molecules encountered by outgoing radiation falls off. At some point, the atmosphere will become transparent to upward-traveling radiation, and that is the point where the radiation will be emitted out into space.

The key effect then is this: if there is more CO_2 in the atmosphere, the point at which the atmosphere becomes transparent to thermal radiation will move to a progressively higher altitude. But since the temperature at higher altitudes is colder, the thermal radiation emission power at this altitude will decrease (recall that the emission rate goes as the fourth power of the temperature at the point the atmosphere becomes transparent). Less radiation emitted outward means there will be an imbalance in outgoing radiation versus incoming radiation, and Earth will heat up. Eventually, a new temperature equilibrium will be established.

Thus, Ångström's concern about the possible saturation of absorption by CO_2 was misplaced, independent of the fact that the details were also incorrect. What is important is that the atmosphere is not a single slab, but that the temperature and density vary with altitude.

My use of the term "blanket" in the last chapter was more than merely poetic. The temperature at the outer surface of an insulating blanket is colder than at the inside because of the insulating material. Thus, the blanket radiates away less of your body heat. The inner surface radiates back downward at a higher temperature, keeping you warm. That is precisely what happens with the atmosphere.

The major difference between this analogy and the actual process for the atmosphere is that for the blanket, the source of heat resides inside the blanket (your body), while for the atmosphere, the source of heat is external—the sun, whose rays make it successfully downward through the atmosphere to heat up the Earth and allow it to radiate upward, like your body, into the atmospheric blanket. (But, of course, the other major difference is the fact that the blanket traps the air inside, like a greenhouse does, so convection and conduction cannot play a role as they do on Earth.)

Because this aspect of CO_2 absorption is both qualitatively and quantitatively important, it is worth putting some meat on the above discussion by showing actual data and the results of some detailed atmospheric modeling predictions.

Shown here are two coincident measurements of infrared radiation emission by the atmosphere in a cloud-free evening at the arctic ice sheet, compiled by researchers at the University of Wisconsin. The upper image is the infrared emission received by a detector located 20 km above the ice sheet, looking *downward*, while the lower image displays the radiation spectrum received by a detector on the ground looking *upward*. To guide the eye, superimposed on the actual data are dashed lines that display the radiation spectrum one would predict for a perfect black body radiating at the labeled temperatures (Figure 6.5).

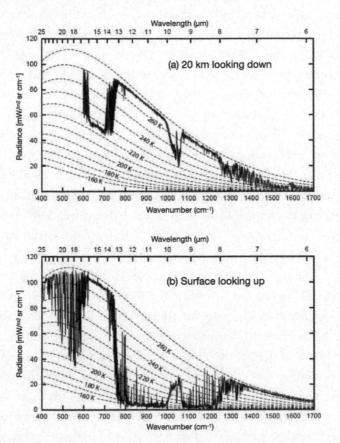

Figure 6.5[29]

What the detector looking downward observes in the frequency band between about 600–700 (in the units used here, which are 1 over the wavelength of the radiation, in cm) is a dip in the spectrum. This is due to absorption by CO_2 in the atmosphere of the radiation that would otherwise come from the surface of the Earth, which in the temperature units here is about 270° Kelvin, or just below freezing. Notice that above this dip, where the atmosphere is largely transparent to infrared radiation, the emission curve follows precisely the radiation pattern one would expect from a hot body at the temperature of the Earth's surface. In the dip, however, the radiation emission is characteristic of a body *at a much colder temperature*, around 225°K, or about -46°C (-51°F). This is because this radiation is coming from a higher altitude, at the point where the atmosphere becomes transparent to upward-going infrared radiation. Because the characteristic temperature at this altitude is much colder, the intensity of radiation emitted in this frequency range is much smaller. In other words, the radiative power emitted by Earth in this frequency band is much less than it would be if the atmosphere were transparent in this range. This reduction in emission causes the blanketing effect I described. Radiation in this region is trapped and reradiated down toward the Earth's surface, keeping it warm, and not out into space.

To demonstrate that this is the case, look at the lower curve. Radiated downward toward the surface in this same frequency band, there is much more radiation received than received above the band, where the atmosphere is transparent to the infrared. In this band the atmosphere is opaque to infrared radiation, so the radiation that makes it down to the Earth emanates from a region much closer to the surface, which is at a much higher temperature (like the inside of the blanket in my blanket analogy). This keeps the surface warmer than it would otherwise be.

The data therefore validates the theoretical prediction.

Now to the models, which themselves are empirically based. Using the best available laboratory data on the infrared absorption properties of CO_2 as a function of frequency, numerical codes incorporate this data with models of the atmosphere to predict radiative emission from the Earth. Comparing the predictions and the data tells us how good the models are likely to be.

One can do more with a good model, however. One can obtain a prediction for what might be expected for a hypothetical world with a CO_2 concentration that differs from the present concentration. Figure 6.6 shows a prediction for what a hypothetical satellite located 20 km above a warmer spot (corresponding to the Earth's surface average temperature of about 15°C or about 60°F) at the Earth's surface, looking downward.

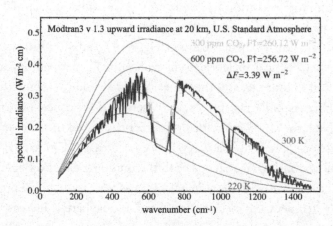

Figure 6.6[30]

The lighter curve represents the prediction for a CO_2 abundance of about 300 ppm, just at the time when Arrhenius and

Ångström had their debate in 1900. The darker curve shows the prediction for a CO_2 abundance of double that value. Notice that as predicted in our earlier discussion, the wings of the dip in the emission spectrum are wider for the higher CO_2 abundance, resulting in increased CO_2 absorption in the atmosphere. As a result, the atmosphere radiates into space at a slightly higher and colder elevation. The effect is less dramatic than that showed in my earlier sketches describing the impact of saturation, however. Indeed, the difference in radiative power emitted upward for a CO_2 abundance of 600 ppm is only about 1 percent less than that emitted for 300 ppm. The difference, about 3.4 watts per square meter, represents a decrease in power radiated away by the Earth compared to the power received by the sun. This decrease in emission will ultimately heat both the ground and the atmosphere until both achieve new higher temperatures. This will emit more radiation back out into space and a new equilibrium balance will once again be obtained.

This is shown schematically in Figure 6.7, adapted from an American Chemical Society presentation, showing temperature as a function of altitude, and the intensity and altitude at which the atmosphere is radiating into space. When the CO_2 concentration is doubled, the altitude of emission is increased, and the intensity of emission (length of the wavy line) is decreased. Ultimately, a new energy balance is reached with the outgoing emission being the same as it was with the original CO_2 concentration. In the process, the temperature of both Earth and atmosphere is increased as a function of altitude (solid vs. dashed line).

This 1 percent difference of 3.4 W/m^2 in emitted radiation due to increased CO_2 atmospheric concentration corresponds to the additional amount the atmosphere will radiate back down toward Earth, in what has become known as "radiative forcing."

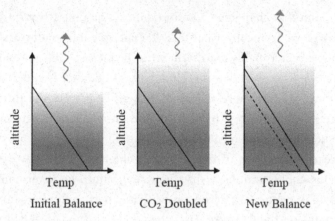

Figure 6.7[31]

Calculating radiative forcing has become the standard tool for quantifying the changing IR absorption in the atmosphere and its effect on changing the energy balance between incoming radiation from the sun and outgoing radiation from Earth. A positive value of radiative forcing means less energy escapes from the atmosphere to space, and as a result there is a positive imbalance in incoming solar radiation versus outgoing IR radiation, which will cause the Earth's surface temperature to rise.

Radiative forcing is the most important single quantity parameterizing the response of Earth's climate to the presence of humanity. Essentially all of the predicted impact on Earth's climate due to an increase in CO_2 and other greenhouse gas abundances comes down to determining a detailed, precise, and accurate estimate of forcing, expressed in terms of additional heat power deposited in the atmosphere-Earth system compared to that radiated to space, in Watts/m^2.

Before turning to our modern understanding of this quantity and its expected impact on Earth's temperature and climate, it is interesting to return to Ångström and Koch's original experimental claim that changing CO_2 abundance wouldn't affect infrared absorption. Ironically, Koch's 1900 experimental result turns out to have been roughly consistent with the picture I have presented here. We have seen that doubling CO_2 abundance is expected to produce slightly more than a 1 percent change in the infrared heat power absorbed by CO_2 in the atmosphere. Koch varied the net CO_2 abundance in his apparatus by only a factor of 1/3. A recent reanalysis of the predicted change in CO_2 absorption for his configuration versus the actual values he determined is shown in Figure 6.8, for an equivalent experiment where instead of reducing the CO_2 pressure, the length of the CO_2-filled tube traversed by the IR radiation is reduced from 30 to 20 cm.

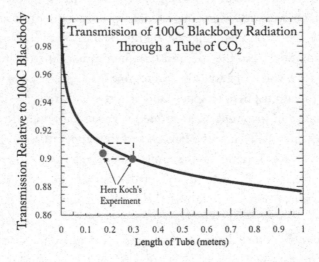

Figure 6.8[32]

As can be seen, the absorption measured by Koch for the 30 cm tube is in good agreement with the predicted absorption in CO_2. Moreover, the predicted change in absorption when the length of the tube is reduced to 20 cm (or the pressure is reduced by 2/3) is less than 1 percent. Koch reported an increase in transmission for the lower pressure of "not more than 0.4 percent." If he overestimated the accuracy of his claim by a small amount, this would make his lower datapoint consistent with the correct value. Since his boss was convinced there would be no effect, observing a change that might have been as large as 1 percent was, for Koch, more likely to be consistent with no change at all. The moral of this story is that laboratory assistants have to be very careful not to be influenced by the force of their bosses' biases!

CHAPTER 7

BELIEVE IT OR NOT

I favor the open mind.
I certainly do not advocate that the mind
should be so open that the brains fall out.
—ARTHUR HAYS SULZBERGER

THE SIGN OF A GOOD THEORY IS WHEN MODELS BASED ON IT GET more detailed and more precise theoretical predictions based on the improved models then agree better with the data. This is the case with the original greenhouse effect, whose existence and theoretical interpretation are by now essentially unimpeachable. The physical principles behind the calculation of radiative forcing—the chief driver of climate change—are the same as those that underlie the two-hundred-year-old explanation of why the Earth's surface is 33°C warmer than it would be without its current atmosphere.

When discussions turn to the issue of global warming, it is easy to lose the forest for the trees. Debates can focus on the detailed model assumptions made to calculate radiative forcing terms and their subsequent impact on climate. Each successive complexity that is included, from the detailed structure of the atmosphere, both vertically and across the Earth's surface, to the role of convection, the importance of clouds, ocean mixing and currents, pollution, Earth's albedo, and so on, is sometimes portrayed as a possible new Achilles heel.

But what should be the most striking takeaway, beyond the fact that ever-more-detailed models generally fit the data better, is the fact that the basic physics captured in even the simplest models gives answers that are consistent with the data. This suggests that they capture the fundamental physics, which is sound.

One could, however, argue that this is somehow an accident—that a conspiracy of competing alternate causes and effects ends up yielding a fortuitous agreement with measurements so our interpretation of the current agreement is both qualitatively and quantitatively wrong. But this seems increasingly without foundation. The evidence, both theoretical and experimental, that I am aware of suggests that details now being debated among experts in climate modeling will not dramatically alter our understanding of the key scientific principles driving climate change in the current era.

I liken the situation to the design of ultra-performance sports cars and the physics of internal combustion engines. The fundamental physics behind internal combustion engines is straightforward. Using this, one can understand roughly how much fuel is needed to make an internal combustion engine with a certain number of cylinders to power a car of a known weight up to a given speed. This is something we could work out in principle in an undergraduate physics class.

At the same time, huge amounts of money and time are spent by the world's best engineers to develop more efficient and faster cars, planes, trains, and so on. Complex modeling is necessary to tweak automobile performance so such things as turbulence within the engine and the air surrounding a speeding vehicle, expansion and feedback between different parts of the engine mechanism and the vehicle chassis, and other detailed interactions are taken into account. The modeling is so complex that it would be impossible to explain it accurately without explicitly going through all these intricacies. But in spite of the fact that these details and more may be necessary to optimize the behavior of the vehicle, it is irrational to suppose that a better understanding of these details is likely to demonstrate that the fundamental picture, including the original physics of internal combustion engines, needs to be revised.

In any area of physics where the fundamental principles are well established, and many independent measurements confirm the same picture, it is improbable that more sophisticated models, which may nevertheless illuminate new features, will force us to throw out the baby with the bathwater.

·····

Let's return then to Arrhenius, who, while having ingeniously accessed rough and incomplete data, built a model based on well-understood radiation physics and derived a general "rule" for associating CO_2 abundance in the atmosphere with the global surface temperature of Earth. His rule, based on the available data and the physics at hand, implied a "logarithmic" relationship between Earth's surface temperature and the atmospheric CO_2 concentration. As I have described, this is the more modern way of framing his statement that the surface temperature will vary by regular linear increments if the CO_2 concentration is increased by constant multiplicative factors.

Over a hundred years later, using a much more sophisticated understanding of the theory behind infrared absorption and the best data on CO_2 spectral absorption, as well as new numerical models of the atmosphere, the functional relationship that climate modelers derive is essentially the same, up to a proportionality factor that is about two times smaller. That Arrhenius got that close using very rudimentary data may seem surprising. But since the fundamental physics associated with thermal radiation has not changed significantly, perhaps it is not as unexpected as it might otherwise seem.

Given that absorption by CO_2 directly affects the radiation balance at the top of the atmosphere, it is simpler and more direct to derive a functional relationship not between CO_2 abundance and temperature, as Arrhenius attempted, but rather between CO_2 abundance and radiative forcing. That is how things are generally expressed today. Recall that radiative forcing is defined as the change in total radiation power (in watts/m²) emitted by Earth into space as some quantity, like the CO_2 abundance, is changed. Once one determines the radiative forcing terms, one can then use atmospheric and climate models to determine the predicted change in temperature at the Earth's surface, as well as other climate changes.

The sophisticated computer models that calculate radiative transfer through many layers of atmosphere from the ground up to produce the figures in the last chapter, which agreed well with observations, yield the following relationship between CO_2 and radiative forcing:

$$\text{Radiative Forcing } (\Delta P) = 5.35 \ (\pm 0.5) \ \ln(C/C_0) \ W/m^2$$

Since it is fundamental to the basis of modern climate science, I felt it useful to present it explicitly. Here, the radiative forcing (change in power radiated into space) is with a CO_2 concentration, C, calculated compared to a baseline value, which by convention is the value in 1750, at the beginning of the industrial revolution, when $C_0 \approx 278$ ppm.

If one were to take $C_0 = 300$ ppm, the value appropriate to the time when Arrhenius made his predictions, then the range of predicted forcing when $C = 600$ ppm would be 3.36–4.05 W/m². How does this compare with Arrhenius's prediction? To answer this question, we need to be able to convert radiative forcing into a surface temperature change. This is not so straightforward, and it is not clear that the same conversion factor would prevail for different CO_2 and temperature values due to a host of other effects that can enter in as temperatures increase, from increased water vapor pressure in the atmosphere to effects of convection, clouds, surface albedo, heat storage in the oceans, and temperature variation across the globe.

Nevertheless, one can make a simple approximate estimate based on the fact that a reduction in power radiated into space leads to that power being stored in radiant heat in the atmosphere. Assume that this radiant heat makes its way down to the Earth, which then heats up and radiates it back. Since radiated power goes as the fourth power of temperature, for small changes in the total radiated power one can then estimate that the change in temperature compared to the pre-existing temperature would be approximately 1/4 of the ratio of the radiative forcing compared to the original total radiated power.

Assuming a radiation forcing of 3.7 W/m², in the middle of the predicted range, and that the original radiated power from the Earth and atmosphere into space at equilibrium is about 245 W/m², then we would find that the predicted change in surface temperature for a doubling of CO_2 would be about 1°C. Another way of expressing this is that for small changes in radiated power by the Earth, the factor relating radiative forcing in W/m² to change in surface temperature in degrees Celsius, using this very simple approximation, is about 0.3.

It turns out that this is a conservative underestimate. When the other factors mentioned above are incorporated into more sophisticated models, the roughly linear factor relating the predicted change in average surface temperature to the predicted radiative forcing is about 0.7–0.8, depending on the models used. This suggests the mean temperature change for the surface of Earth, if the CO_2 concentration doubles compared to the 1900 value, would be in the range of 2.5–3.5°C, about 2.5–3 times larger than the naïve prediction based simply on scaling radiative power versus temperature, and about 2 times smaller than the temperature change predicted by Arrhenius. While the variation between these estimates is not negligible, nevertheless at the order-of-magnitude level, these estimates are all relatively close, predicting a temperature change on the order of 1 percent, and not 0.1 percent or 10 percent of the original surface temperature.

Fortunately, or perhaps unfortunately, we can actually empirically check this estimate with at least one data point. We can compare the mean surface temperature of the Earth today versus that in 1900, given that the CO_2 abundance in the atmosphere has changed from 300 ppm to about 415 ppm. Using the modern estimates above, the mean prediction for the temperature change over that period, ignoring other effects for the moment, is about 1.3°C. Figure 7.1 shows the data from climate.gov.

Global Land and Ocean
January–December Temperature Anomalies

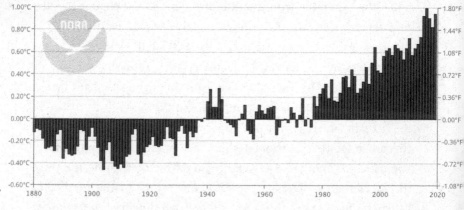

Figure 7.1[33]

As can be seen, the actual change, accounting for the year-to-year fluctuations is in the range 1.15–1.35°C. That is pretty good agreement with the prediction.

In fact, the agreement is actually better than one might have expected because, as I indicated, the prediction I presented assumes just a change in CO_2 abundance without considering other possible radiative forcing factors present, such as additional greenhouse gases, surface albedo, clouds, solar irradiance, or pollution. Some of these, such as increased cloud cover or conversion of forests into urban areas, produce a negative forcing, whereas others, including contributions from other greenhouse gases such as methane and other hydrocarbons, produce a positive forcing.

In 2007 the UN Intergovernmental Panel on Climate Change (IPCC) released a chart showing the best estimates and the predicted uncertainties for various radiative forcing factors for 2005 compared to 1750. The largest relative uncertainties involved the effect of clouds and aerosols in the atmosphere, while the predicted uncertainty associated with CO_2 abundance was relatively small. When combined, the average predicted radiative forcing at that time was about 1.6 W/m². This is about 5 percent smaller than that predicted for CO_2 alone, confirming that the estimate we obtained by just considering CO_2 was a reasonably good approximation to the full situation.

Alas, humanity has not stood still since 2005. Greenhouse gas emission has continued to rise. We remain on track to double the CO_2 concentration since 1750 in this century. Figure 7.2 is a chart showing the forcing due to greenhouse gases since 1980.

At the same time, uncertainties in various other factors, including impact of aerosols and clouds, have decreased. In 2013, the IPCC updated their chart, also listing the increase in net anthropogenic radiative forcing over the years. Thus, the net

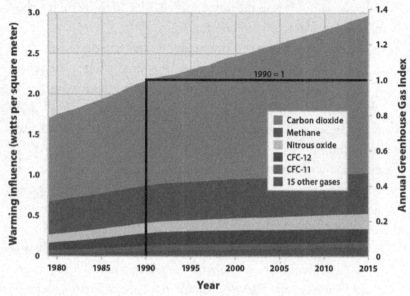

Figure 7.2[34]

forcing in 2011 was estimated to have a mean value of 2.29 W/m², a 40 percent increase compared to the 2005 estimate, and almost double that estimated from 1980. The chart is shown in Figure 7.3.

For some of you, the last three chapters may have been something of a slog, so I applaud you for making it this far. My reason for presenting the fundamental theoretical framework and the experimental tests of that theory in some detail here has been twofold. First, the basic physics is relatively straightforward and can and should be explained comprehensibly to interested bystanders—which should include basically most of humanity—following to some extent the scientific developments that led scientists to where we are today.

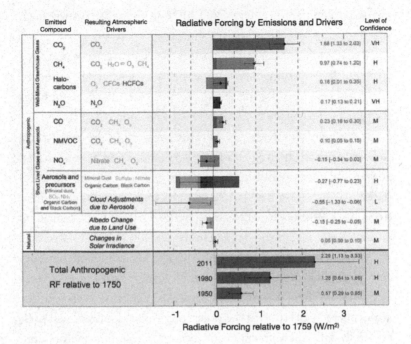

Figure 7.3[35]

Second, the predictions presented here represent the baseline perspective from which the discussions of the rest of this book should be viewed and from which each of us can ultimately draw conclusions about the possible risks, impacts, and necessity for global action or inaction on this issue. I therefore felt I owed it to you to give you the most complete and coherent general perspective I could. The physics is clear. If it isn't to you, or still seems sketchy, blame my presentation, not the science.

Before concluding this preparation for what follows, I will add one more bit of data that helps confirm that estimates of greenhouse-induced temperature change are empirically reliable. I have discussed essentially one datapoint so far: the predicted change in temperature between 1900 and today versus the measured change. We can do better than that, however.

Recall that ice cores have allowed us to measure the CO_2 content of the atmosphere going back at least eight hundred thousand years by looking at the composition of the gas bubbles trapped in the ice. Ice cores allow us to do more than this.

Ice is made of water, H_2O. However, there are two stable isotopes of both hydrogen and oxygen. Heavy hydrogen, called deuterium, exists with a ratio of about one part in ten thousand in the oceans. ^{18}O, a heavy isotope of oxygen containing ten neutrons instead of eight neutrons in the oxygen nucleus, makes up about 0.2 percent of the oxygen in the world's oceans on average.

Now, because deuterium and ^{18}O are heavier than the more common versions of hydrogen and oxygen, it takes more energy to evaporate them out of the oceans. Also, as water vapor in the atmosphere heads toward the poles, the heavier isotopes will preferentially fall when precipitation occurs. These two factors contribute to what is called "fractionation," and both are temperature dependent, with more fractionation during colder periods. Thus, if one measures the ratio of deuterium to hydrogen, and ^{18}O to ^{16}O in ice cores, during colder times we will find systematically greater depletion of these heavy isotopes in the ice core samples, and during warmer times systematically less depletion. Thus, ice cores give us geological record not just of CO_2 abundances, but also atmospheric temperatures! Figure 7.4 is a plot from the Vostok ice core in Antarctica, going back over four hundred thousand years, showing CO_2 abundances and inferred arctic temperatures compared to the average temperature from 1961 to 1990.

It is hard not to be struck by the correlation, which continues back at least four hundred thousand more years using data from the Dome C ice core in Antarctica (Figure 7.5).

Vostok Ice Core Record:
Carbon Dioxide versus Temperature.

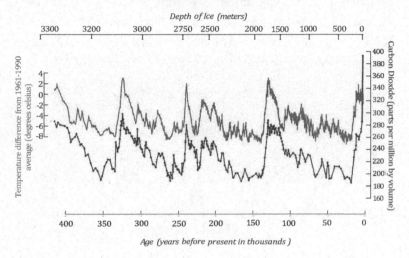

Figure 7.4[36]

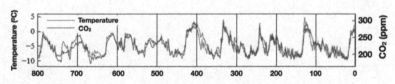

Figure 7.5[37]

During this period, long-term periodic changes in the Earth's orbit are what initiated the climate change. But without the feedback due to the radiative forcing induced by the resulting changing CO_2 concentration and the changing surface albedo as ice sheet size changed, the actual temperature changes would have been far smaller.

Using the relation derived earlier for the predicted change in temperature (°C) as ≈ 0.75 times the total radiative forcing, along with the predicted relationship between CO_2 abundance and its contribution to forcing (about one third of the total forcing in this case), Jim Hansen compared the predicted and observed temperature change for the Vostok 440,000-year core (temperature changes divided by two to fit on the plot) in Figure 7.6.

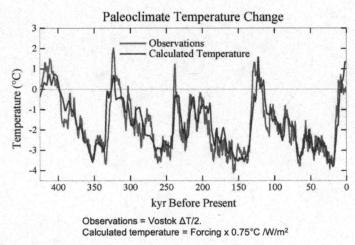

Observations = Vostok ΔT/2.
Calculated temperature = Forcing x 0.75°C /W/m²

Figure 7.6[38]

That this relatively simple set of approximations to the detailed model results provides as good a fit as this provides, at least to me, strong further confirmation that the physical arguments and predictions associated with radiative forcing and CO_2 abundance are sound.

CHAPTER 8
SOME LIKE IT HOT

Assure me that I yet may change these shadows
you have shown me.
—CHARLES DICKENS, *A Christmas Carol*

I HAVE OFTEN SAID THAT I ONLY MAKE PREDICTIONS ABOUT TWO trillion years *into* the future. First, the cosmos will be colder, darker, and simpler. Second, no one will be around to check if I am wrong. Predicting instead what might happen in the coming century is fraught with inevitable uncertainties, as is humanity. However, the laws of physics do put some restrictions on tomorrow based on what we know about today.

In this chapter I want to summarize some future predictions oddly reminiscent of Dickens's *A Christmas Carol*. Christmas Past couldn't be changed, and we cannot change our past, neither can we change some of the present and future consequences of humanity's past actions. Christmas Present invited Scrooge to begin to imagine how actions he took that day might pan out. And Christmas Future, the most terrifying of all, invited Scrooge to look at the consequences of the future as it *might be* with an invitation to ensure that future never happens.

We often hear or read claims about our future with global warming, but these predictions range from the guaranteed to the likely, the plausible, and the possible, often with few distinctions made. There can be a world of difference, as Dickens devoted a whole story to demonstrate, between the future that will be and the future that might be. I want to keep track of those differences here.

I will begin with the guaranteed developments because they are already built into the system, and simple physics arguments constrain what will happen next.

First, what happens if we stop emitting tomorrow? I am tempted to suggest a COVID-19 pandemic-related cause, as that seems to be central on everyone's minds as I am writing this, but I sincerely hope that by the time this book appears there might be other things to talk and write about.

Be that as it may, Earth's surface temperature has already changed by about a full degree in the past century as a result of the emission of about 450 gigatons of carbon into the atmosphere, as well as other forcings. If carbon production stops, will the temperature return to its earlier value? Not anytime soon. As demonstrated in 2008 by Susan Solomon and colleagues, the time period over which it will drop by half, due to varying rates in ocean uptake, exceeds one thousand years. Much of what we have put up there will stay there till about the year 3000. That is, unless we figure out some new technology to remove it, or an effort at geoengineering that manipulates the atmosphere on a global scale to otherwise remove the radiative forcing that already exists.

Solomon *et al.* considered numerical climate models where CO_2 emissions continued at a 2 percent yearly increase after 2008, reaching CO_2 atmospheric concentrations ranging between 450–1200 ppm, before emissions abruptly ceased. Figure 8.1 was their result.

The CO_2 abundance "quickly" drops by about 20 percent over a century or so, due largely to some rapid CO_2 uptake by the terrestrial biosphere, which it then begins to give back a century later. After that, the abundance falls much more slowly, and

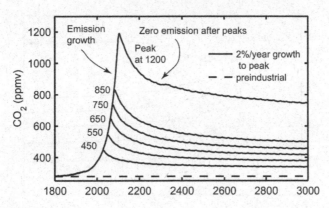

Figure 8.1[39]

a millennium later they find that 40 percent of the original CO_2 peak abundance on top of the original value of 280 ppm is still in the atmosphere. Similar 2010 calculations by a Canadian group who considered two cases—an abrupt end to CO_2 emission in 2010 or an abrupt end in 2100 with a peak concentration of almost 800 ppm—found similar results, with 55 percent of the peak CO_2 concentration remaining after a millennium.

While calculating CO_2 abundance evolution is more direct, converting this into a global surface temperature change involves more model-dependent assumptions. Nevertheless, the models allow one to numerically estimate the mean global temperature evolution, including fluctuations from year to year, as well as the distribution of temperatures across the globe. As we saw, the global temperature predictions appear to be accurate thus far to about 0.1–0.2 degrees Celsius. Figure 8.2 is a result for global average surface warming from the Solomon study.

The key takeaway here is that the relatively recent rapid temperature rise that has occurred as long as CO_2 emissions have continued unabated over the past fifty years *ends very quickly*

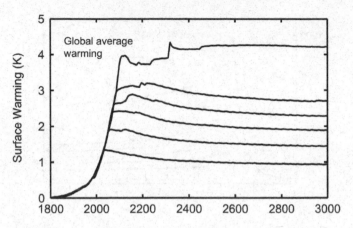

Figure 8.2[40]

upon cessation, but the temperature itself does not noticeably decrease even as CO_2 levels themselves slowly decrease over the next thousand years. This is due in large part to a factor that will play a key role in much of what follows and in much of the mid- and long-term future of the planet. The deep oceans take a long time to mix and equilibrate the additional heat that has been pouring into the Earth due to the greenhouse effect. Hence the oceans will continue to warm for many hundreds of years after the source of excess heat has disappeared. Land masses, on the other hand, store less heat and therefore cool more readily as the radiative forcing decreases.

For this reason, since the Southern Hemisphere has much more ocean and less land than the Northern Hemisphere, there is expected to be a differential temperature change across the planet's surface, and that change will itself change with time. Figure 8.3, for example, displays the relative temperature changes across the planet between 1850 and 2105 in the Canadian study where carbon emission continues unabated until 2100, and below it the subsequent temperature change between 2105 and 2995.

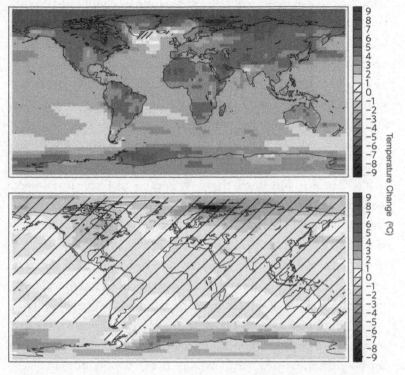

Figure 8.3[41]

As can be seen in the top figure, the highest 'immediate' rise is in the Northern Arctic, where a shallow ocean is surrounded by large land masses. Hundreds of years later, in the lower figure, the Arctic will have already begun to cool significantly whereas the Southern Ocean and Antarctica continue to warm well after CO_2 stopped rising. (Similar but smaller changes occur in the model in which emission stops in 2010. We are somewhere in between, having had one additional decade of CO_2 emission.)

This can be understood in large part because of the huge latent heat stored in the deep oceans, which store over 90 percent of the heat that has been added thus far to the climate system, and which, as I have indicated, takes some time to equilibrate. In the Canadian study, an estimate of the temperature of the Southern Ocean at a high latitude of 66°S found ocean temperatures continuing to increase throughout the next millennium by as much as an additional 3°C in the pessimistic case where carbon emissions don't cease until 2100, ultimately slightly exceeding the average global surface temperature rise throughout this period.

Their overly optimistic scenario, in which carbon emissions ceased in 2010, is worth focusing on because it describes the future as it *will be* (actually it will be worse because as I have just noted, we have continued to emit CO_2 in this decade). *It yields a further rise in ocean temperature through the next millennium by almost 0.5°C, an amount that is essentially already written in stone because that heat has already been deposited over the past century.* This will have inevitable consequences for the planet, some of which I will discuss in more detail in the next chapter.

To get an idea of the heat capacity of the world's oceans compared to the heat dumped in the oceans by humanity so far, consider the measured rise in average ocean temperature in 2019 of 0.075°C compared to the 1981–2010 average temperature, as recently compiled by an international group of researchers. Just this measured temperature difference—which I remind you does not yet reflect the full ocean response to the additional heating by CO_2 radiative forcing thus far—corresponds to the additional heat that would been produced by setting off 3.6 billion Hiroshima atom-bomb explosions in the ocean or about five Hiroshima bombs of heat every second, day and night, 365 days a year for the past twenty-five years.

Besides the overall change in ocean temperatures, there are two other long-term impacts that have already been written in stone: one direct and one indirect. The first impact is due to the simple thermal expansion of water as its temperature increases, a fundamental basic physics result.

When global warming is discussed, most often reference is made to the predicted rise in sea levels due to the possible continued melting of glaciers and ice shelves in Greenland or Antarctica. However, as I shall describe in the next chapter, the dominant single source of sea level rise in recent times has not been due to this effect but rather to thermal expansion due to the average ocean temperature rise thus far.

Figure 8.4 is a graph of the measured sea-level rise, using satellite data between 1992 and 2015, of about 3.3 mm per year. Fully half of this effect is due to thermal expansion of water and is independent of any uncertainties associated with glacial melting.

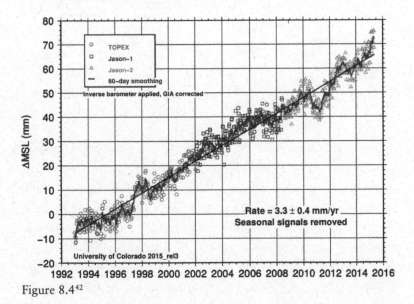

Figure 8.4[42]

As the world's oceans continue to increase in temperature, *even long after global industrial CO2 emission may have ceased,* the ocean level will continue to rise, again independent of glacial melting effects. Figure 8.5 is the prediction from the Solomon simulation discussed above. Once again, note that the lower curve represents what is inevitable, even if we had stop emitting CO_2 in 2010:

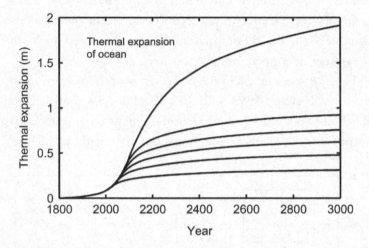

Figure 8.5[43]

Note that the lowest curve occurs with a peak concentration of 450 ppm, close to the current value of 417 ppm. This means a long-term sea level rise of about ¼ of a meter is essentially inevitable given the world's cumulative CO_2 emissions to date, independent of other possible sources of sea level rise, including melting of major ice sheets. If CO_2 emissions continue at more or less their current rate for the rest of this century, the minimum sea-level rise we can expect is about one meter. For this case, the more recent 2010 Canadian model analysis predicts a sea level rise slightly in excess of one meter.

The other more indirect impact that is likely to continue well after our current CO_2 emission rate declines or ceases involves a predicted change in the precipitation patterns around the world with significant potential regional impacts. These are due to predicted changes in global ocean currents affected in part by differential global temperature shifts, deep mixing of heat and increased fresh-water content due to glacial melting.

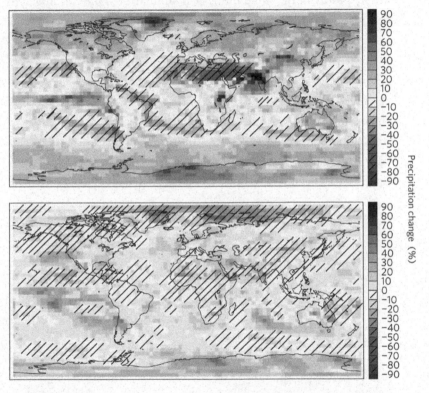

Figure 8.6[44]

A detailed discussion of models of ocean dynamics and currents is beyond the scope of this book, but I will mention one predicted long-term impact because it follows almost directly from a direct prediction I have already discussed: the differential long-term heating of the Southern Hemisphere versus a cooling of the Northern Hemisphere after emissions cease. The net impact of this temperature shift is predicted to produce a consequent precipitation change: continued drying in the low latitudes of North Africa, South America, and Australia.

As for the global temperature pattern predictions displayed earlier, and from the same numerical simulations, the top image shown is the predicted change in precipitation between 1855 and 2105 if emissions continue till 2100. The lower plate displays the predicted change between 2105 and 2995 in this case (Figure 8.6).

Solomon and colleagues demonstrated just how severe the impact could be in various locations during dry season in 3000 as a function of the peak CO_2 concentration before cessation (Figure 8.7). The shaded region represents the range of precipitation experienced during major past droughts, such as the "dust bowl."

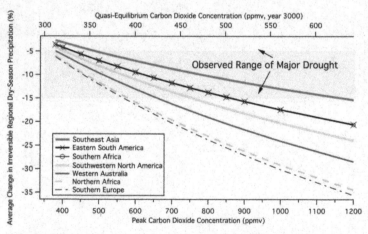

Figure 8.7[45]

The fact that CO_2 will remain in the atmosphere long after humanity stops or reduces its global emissions, combined with the fact that the potentially negative observable consequences of its current abundance will in some cases take centuries to fully manifest themselves, should affect our thinking about the consequences of continuing business as usual and the potential need for global action sooner rather than later.

Once again, I remind you that pre-industrial carbon abundance in the atmosphere was about 600 billion tons. We have added about 500 billion tons to that amount to date. If emissions continue at the current rate till 2100, we will have added by that time an additional 2.2 trillion tons to the atmosphere.

Say then, for example, we wish to limit global temperature rise to some specific value, perhaps 1.5°C. Then, because the cumulative CO_2 abundance in the atmosphere simply increases each year we add more CO_2 and doesn't significantly decrease if we stop emitting, we simply have to ask how many more years we can keep emitting at the current rate before the net radiative forcing by CO_2 produces the predicted temperature change. For 1.5°C we are already 90 percent of the way there, meaning in ten years we will have already passed that threshold if current emission rates remain constant.

Let's say we relax our goal and accept a temperature rise of at most 2°C, the target of the 2015 Paris Agreement. One of the arguments that has been used against trying to moderate emissions to meet this goal is based on the claimed economic costs of trying to achieve it. I won't debate that issue here, even though I think it is spurious, but rather talk about the science. And here the science presents a dilemma.

Each year we delay trying to limit the greenhouse temperature rise to be less than some target, it becomes harder and more costly to achieve that goal. How much would one have to reduce emissions each year so that the total cumulative emission is less than an amount required to yield a two-thirds probability of global temperature rise being less than 2°C? Figure 8.8 is a chart from 2009, called the "ski slope" diagram, showing what would have been required if we had started this process globally in 2011 (beginner slope), 2015 (intermediate slope), or in 2020 (expert slope):

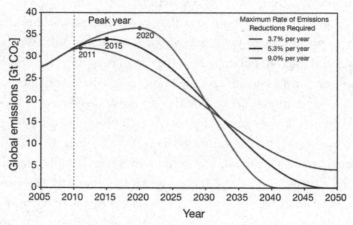

Figure 8.8[46]

Observe that in 2011 we would have only had to consider a 3.7 percent annual reduction in emissions. Now we would have to somehow arrange a 9.0 percent annual reduction.

Figure 8.9 is a similar graph, projecting things further into the future, also showing actual emission rates over the past forty years. You can see that even if emissions don't keep growing, but also don't start falling, then by 2030 or so we would essentially have to turn off the spigot, which would be virtually impossible.

One can compare these hypothetical pathways with the actual range of likely trajectories given current global policies. One of these futures will occur. We might get to choose which one—if we actually think about it enough to make the choice (Figure 8.10).

CO₂ reductions needed to keep global temperature rise below 2°C

Annual emissions of carbon dioxide under various mitigation scenarios to keep global average temperature rise below 2°C. Scenarios are based on the CO₂ reductions necessary if mitigation had started – with global emissions peaking and quickly reducing – in the given year.

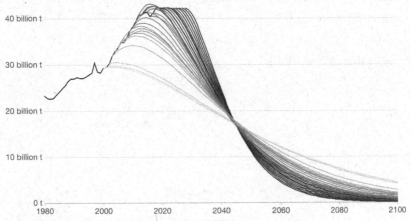

Source: Robbie Andrews (2019); based on Global Carbon Project & IPPC SR15
Note: Carbon budgets are based on a >66% chance of staying below 2°C from the IPCC's SR15 Report.
OurWorldInData.org/co2-and-other-greenhouse-gas-emissions • CC BY

Figure 8.9[47]

Global greenhouse gas emissions and warming scenarios

- Each pathway comes with uncertainty, marked by the shading from low to high emissions under each scenario.
- Warming refers to the expected global temperature rise by 2100, relative to pre-industrial temperatures.

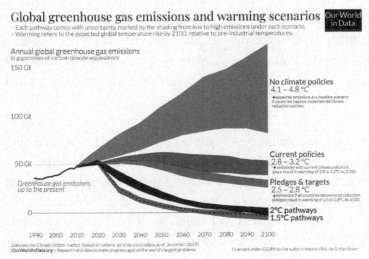

Annual global greenhouse gas emissions
in gigatonnes of carbon dioxide-equivalents

Data source: Climate Action Tracker (based on national policies and pledges as of December 2019)
OurWorldInData.org - Research and data to make progress against the world's largest problems
Licensed under: CC-BY by the authors Hannah Ritchie & Max Roser.

Figure 8.10[48]

Finally, we can extend the emission curve out to yet longer times. Figure 8.11 shows what the range of cumulative human emissions looks like, from business as usual to the current and perhaps unrealizable global pledge of 2.6–3.2°C warming. (Again, remember the pre-industrial CO_2 abundance in the atmosphere was 600 Gt.)

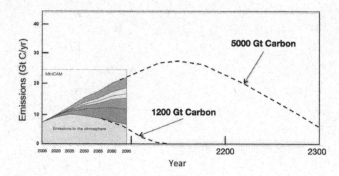

Figure 8.11[49]

Returning to a theme from earlier in this book, the numbers demonstrate convincingly that human emissions are *not* insignificant on a global geological scale. In two hundred years humans will have been able to emit at the very least twice as much CO_2 as existed in the Earth's atmosphere over much of the geological record.

These present a minimalist view of the future as it will be, one way or another, depending on our actions. I have studiously avoided discussing how the increases I have described here will impact on humanity, and also about possible feedback or amplifying mechanisms that might exacerbate these impacts.

In the next chapter I will focus on one particularly disturbing aspect of the future as it *might be*. In so doing, I will display the first graph I saw that convinced me that globally the risks are such that we would be wise to address climate change now.

CHAPTER 9

WHAT WILL BE, WILL BE?

I know now: what is is all that matters.
Not the thing you know is meant to be...

AUGUSTEN BURROUGHS, *Lust and Wonder*

MY MOTHER SANG THE SONG "QUE SERA, SERA (WHATEVER Will Be, Will Be)" to me when I was a young boy, and I found it reassuring because it implied the future was beyond our ken. As an adult, I don't buy it. One of the purposes of science is to predict the future.

In my professional work I have generally focused, with some exceptions, either on making predictions about experimental systems that are so well controlled that the implications of fundamental theory are directly discernible from observations, or I have made predictions about the universe on its largest observable scales. On these scales, fortunately, the universe can be modeled very simply.

It is far harder to model the full complexity of daily life on Earth, much less to predict what will happen tomorrow. But when lives depend on it, we can rise to the task. After centuries of effort, with help from modern satellites, supercomputers, and weather stations across the globe, meteorologists can now accurately predict the weather, at least on a statistical basis, up to ten days in advance. Storms and hurricanes can be anticipated with a great savings in human life.

Weather is different than climate, however. Weather takes place in a single place, at a single time. Climate is longer term, and regional or global rather than local. On global scales, the local variability from place to place can become unimportant. Using general principles of work, heat, and energy flow, the global heat balance of the planet has been understood and measured. And the predictive and experimental capabilities of physics and chemistry have been refined to test and confirm the fundamental principles governing the evolution of climate on our planet on its broadest scales. Absorption of infrared radiation by greenhouse gases has been verified, and the energy balance governing the Earth/atmosphere/sun are understood. The most direct prediction involves energy transfer, so we discovered radiative forcing. From there, atmospheric models allowed a prediction of surface temperature changes that could be tested. With temperature changes come a whole new set of phenomena. Sea level rise due to thermal expansion is the most immediate consequence, but we have seen that the impact of the ocean is also significant because it is by far the largest heat sink governing the long-term evolution of climate in response to radiative forcing.

The differential heating of the planet as a function of latitude is the direct consequence of the asymmetric distribution of land mass and ocean as a function of latitude. The enhanced immediate temperature rise of the Arctic is one result, and the long-term heating of the Antarctic will be another. The extent to which this picture is accurate is of vital importance for the future of the planet in the coming millennium because the two greatest reservoirs of fresh water in the form of ice are located at or near the Earth's two poles: in Greenland and Antarctica. And a significant temperature rise in one or both locations could have indirect consequences that may be more dramatic than that due to the direct increase in mean surface temperature alone.

The thermal expansion of water has dominated the measured sea level rise to date, and a continued expansion is basically guaranteed due to the additional heat already stored deep in the oceans. But as has been mentioned, there are other contributions, primarily due to ice sheet disruption in Greenland and Antarctica. While these have been subdominant over recent decades, there is every indication that this will not remain the case. As the National Academy of Sciences put it in a study published in 2011: *The dynamic response of ice sheets to global warming is the largest unknown in the projections of sea level rise over the next century.*

Thus, even if it involves considerations beyond the direct consequences of radiative forcing, it is worth exploring an overview of the data, mechanisms, and predictions for significant disruption of glacial ice and its impact on expected sea level rise.

Glacial melting is accelerating, especially in Greenland. The extent to which this melting will continue to accelerate will depend on a host of factors, and as the National Academy stressed, not all of these are easily estimated direct consequences of radiative forcing. This is, after all, the future *as it might be*.

As far as I am concerned, the most chilling images, if you forgive the pun, come not from the predictive models of the future, but from the data about the past. I remember when I first saw this data over a decade ago—before I had thought about the underlying physics in any depth—I was shocked out of my own complacency and decided there were realistically some serious risks worth worrying about.

Let's start with Greenland. By every standard one can devise, the melting of Greenland's ice sheet is accelerating. Every year seems to top the report from the previous year. The National Oceanic and Atmospheric Administration (NOAA) has issued a yearly Arctic Report Card since 2006, and the National Snow and Ice Data Center (NSIDC) provides regular reporting as well.

First, consider ice mass loss. Already a decade ago studies were demonstrating increasing loss, measured in gigatons/year. Tony Haymet, then director of the Scripps Institution of Oceanography, provided me with Figure 9.1 in 2009, where the Greenland ice mass quantities are shown relative to the ice mass value in 2006.

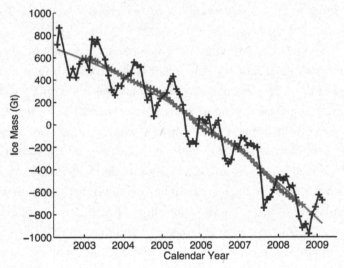

Figure 9.1[50]

The updated Figure 9.2 comes from the 2019 Arctic Report Card, where the timescale is set so that it matches the previous figure on the dates of overlap, and ice mass figures shown are now relative to 2010 and not 2006.

I added one more point to this graphing, coming from two reports in April 2020 from labs that include Columbia University, University of California, and the Jet Propulsion Lab. They

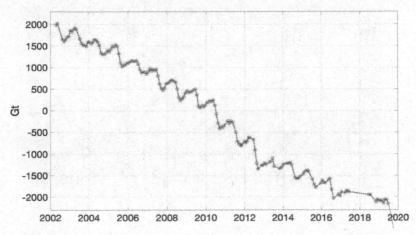

Figure 9.2[51]

reported that over the Arctic summer of 2019, Greenland lost approximately 600 Gt of ice, more than twice the average rate of ice loss, about 270 Gt/yr., between 2012 and 2017. Between 2003 and 2007, by contrast, the average annual mass loss was 180 Gt, and between 2000 and 2003 the rate was 150 Gt/yr. Between 1997 and 2003 the average loss was about 74 Gt/yr., and between 1993 and 1999, before the data shown on these curves, the average mass loss was recorded to be only about 54 Gt/yr. The trend is clear.

Figure 9.3 comes from a satellite that measures the microwave brightness temperature of the surface. Shown below are the annual number of cumulative melt days on the ice. On the left is the annual average from 1979–2007, and the middle is 2012, the most severe melting year until 2019. The image from 2019 includes only the period April–December. The increase in melt in these years compared to the earlier average is immediately visible.

113

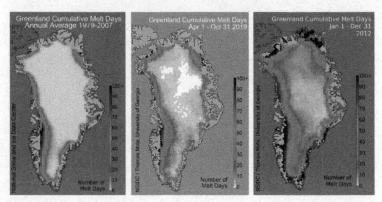

Figure 9.3[52]

One can get another graphical representation of the data by multiplying the total number of cumulative melt days by the area that the melt was detected in. This has been labeled the *standardized melt index*, when the yearly data is compared to the average of the full measurement period from 1979 onward. Figure 9.4 is the index up to 2012.

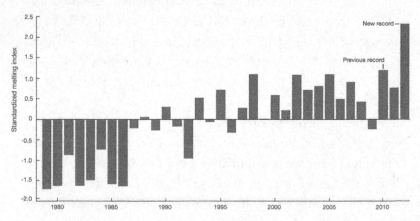

Figure 9.4[53]

This data make it unambiguously clear that the Greenland ice sheet is now melting at an accelerating rate, so Greenland is now contributing an ever-increasing amount to measured sea level rise. Over the entire decade from 1993 to 2003, Greenland ice melt contributed about 0.21 mm to the rising sea level, whereas the 2019 melt in a single year raised sea levels by an order of magnitude more than this.

What can we expect in the long term?

Before we focus on that question, some comments about the interpretation of the current data: First, an acceleration of Greenland's ice melt is to be expected based on the analysis of asymmetric global warming in response to radiative forcing I have already described. We have observed and we expect more initial warming in the far north compared to that in the Southern Hemisphere. Second, it is worth noting that the total ice mass loss in Greenland is not simply due to melting but an almost equal amount in the early 2000s was due to ice discharge, like the calving of glaciers into the sea, and also the interaction of warming ocean water with glaciers in contact with the ocean. These factors are related to yet more complex climate effects. That said, one should note that the ice discharge from Greenland over recent years has not increased noticeably, while melting has. Thus, estimating the total ice mass loss in the future by considering melting alone might not lead to too significant an underestimate of resulting sea level rise.

While the data presented on the previous pages is consistent with what one would expect on the basis of radiative forcing and global temperature increases, a natural skeptic might ask, when presented with this data, about its historical context. Is this anomalous? Is there other evidence of Greenland ice mass as temperatures have varied in the past, and if so, by how much?

Fortunately, there has been a significant amount of data obtained that sheds light on the paleoclimate in Greenland. The results are sobering.

Detailed measurements of marine deposits, coastal terraces, coral deposits, isotope abundances, and ice cores from Greenland make it quite clear that Greenland has been ice-free at times as recently as four hundred thousand years ago, and that the total Greenland ice mass correlates more closely to global temperature than anything else. But, while rapid retreat of glacial ice near the edges at the coasts has occurred in the past over the course of decades, Richard Alley and colleagues concluded in a 2010 *Quaternary Science Review*: "major changes of central regions of the ice sheet are thought to require centuries to millennia."

A number of factors that might have been expected to significantly contribute to ice loss or gain and thus, which might add uncertainty to model estimates, have been shown to be largely absent. Ice loss in response to warming has been relatively insensitive to sea level change, for example. And, while precipitation is predicted to increase with increasing temperature, and this has been documented in increased snowfall in Greenland during periods of warming, that increased snowfall could not compensate for the increased melting and ice discharge, so the net ice mass decreased. As summarized in the 2010 review (*Quaternary Science Reviews*, Vol 29) by Alley and colleagues:

> Paleoclimatic data show that the Greenland Ice Sheet has changed greatly with time. From physical understanding, many environmental factors can force changes in the size of an ice sheet. Comparison of the histories of important forcings and of ice-sheet size implicates cooling as causing ice-sheet growth, warming as causing shrinkage, and sufficiently large warming as causing loss. The evidence for temperature control is clearest for temperatures similar to or warmer than recent values (the last few millennia). Snow accumulation rate is inversely related to

ice-sheet volume (less ice when snowfall is higher), and so is not the leading control on ice-sheet change...sea-level change is not the dominant forcing at least for temperatures similar to or above those of the last few millennia...the limited paleoclimatic data consistently show that short-term and long-term responses to temperature change are in the same direction. The best estimate from paleoclimatic data is thus that warming will shrink the Greenland Ice Sheet, and that warming of a few degrees is sufficient to cause ice-sheet loss.

The loss of the complete Greenland ice sheet would raise sea levels by about 7 meters, and significant changes in the ice sheet have accompanied periods of warming during the six major interglacial periods between ice ages on the planet over the past four hundred thousand years. There is evidence that during at least the warmest period, about four hundred thousand years ago, when local temperatures were at least 2–4°C higher than the present time, that the ice sheet did disappear, with a consequent sea level rise of 7 meters, and during a more recent period, about a hundred thousand years ago when the local temperature was at least briefly in a similar range of 2–4°C above the present time, that sufficient melting occurred to raise sea levels by about 3.5 meters. The uncertainties remain great, however, which should not necessarily be a source of solace. As stated in the abstract of the 2010 study: "The evidence suggests nearly total ice-sheet loss may result from warming of more than a few degrees above mean 20th century values, but this threshold is poorly defined (perhaps as little as 2°C or more than 7°C)."

Guided by this paleoclimate data, Richard Alley and colleagues have modeled the likely longer-term dependence of the Greenland ice sheet on temperature and, using radiative forcing estimates, on the CO_2 abundance. Figure 9.5 shows that various values of the peak CO_2 abundance are assumed, depending on when global emissions might substantially cease.

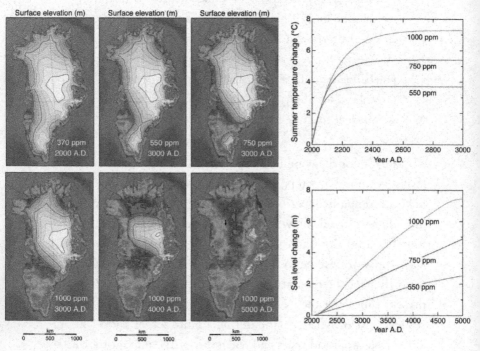

Figure 9.5[54]

In the near term, the mean estimate in 2010 for the guaranteed sea level rise by 2100 due to Greenland ice sheet melting, given a more-or-less certain rise of 2°C before the end of the century, was about 0.05 meters. However, since that time ice mass loss has doubled, reaching the maximum estimated melting rate predicted in models, so a newer estimate closer to 0.1 meters now seems more appropriate. The National Academy quotes an upper bound of 0.16 meters, assuming continued increase in melting combined with a possible increase in ice discharge at the coasts, which currently contributes an equal amount to melting in the total ice mass loss.

As increased precipitation due to increased temperature is now known to not be a significant factor in reducing mass loss, other factors not included in models would tend to increase the actual mass loss rate and would bring them into closer alignment with recent observations. Such factors include recently observed lubrication at the base of glaciers due to large surface melting, which can increase the rate of glacier movement into the sea. And most recently in the 2019 Greenland data, it was observed that much of the extreme melting that year was not directly due to higher temperatures, but rather to high-pressure weather systems in the region that may have been an indirect result of temperature increase. These blocked the formation of clouds, which allowed unfiltered sunlight to melt the snow. In addition, with less snowfall, darkened dirt and soot-covered ice was exposed, which altered the albedo, absorbing more heat than would otherwise be the case. As a result, a Greenland contribution of 0.1–0.16 meters to sea level rise seems a reasonably conservative range to expect this century.

Moving across the globe to Antarctica, the situation is quite different. There, snowfall increases in East Antarctica have on average balanced ice loss in West Antarctica. Again, this is not unexpected on the basis of our earlier predictions, as near-term warming is predicted to be less severe in the high-latitude South, and increased precipitation is predicted there as well.

Nevertheless, over the past two decades the West Antarctic ice sheet has undergone noticeable changes, which I became acutely aware of when I visited in 2015.

The immensity of these icebergs in that region is hard to fathom until you see them up close. When one passes close by your boat, it stands taller than a skyscraper and stretches as far as you can see. I tried to capture one from the bow of our vessel.

55

But these icebergs are small potatoes. A few months before my arrival, in the dark of the Antarctic winter, a giant iceberg, 225 square miles, about ten times the size of Manhattan, had broken off from the Pine Island Glacier on the Amundsen Sea. It was not the first such iceberg, the largest, nor would it be the last. Just this year it shed another smaller clump 120 square miles in area. Since 2012 this single glacier has been shedding fifty-eight gigatons of ice a year, making the biggest contribution to sea level rise, of about 0.1 mm/yr., of any single ice floe on Earth.

These are not the biggest recent icebergs to form off the West Antarctic coastline. A massive ice shelf, called the Larsen Ice Shelf, on the eastern side of the West Antarctic peninsula, opposite the location of the Pine Glacier, has broken off three spectacular state-sized pieces since 2002, with the last and largest, called Larsen C, breaking off in 2017. It weighed one trillion tons and measured about 2,200 square miles, the size of Delaware. The photographs of the rift as it broke off were remarkable.

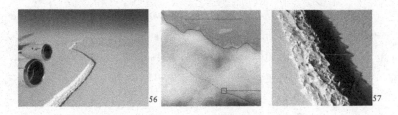

As large as these are, the Larsen shelf calvings do not have any impact on global sea levels because the original ice shelves were already floating on the water. But nevertheless, what they may portend is somewhat more ominous.

While Antarctica has not warmed as extensively as the Arctic over the past decades, it has warmed. Changing wind patterns and related warm deep-ocean currents being pushed southward, as predicted, have warmed West Antarctica most, as seen here with the average temperature change from 1957–2006 displayed (Figure 9.6).

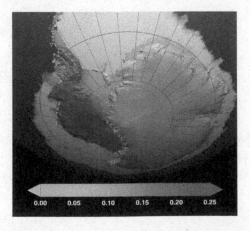

Figure 9.6[58]

This warming trend has accelerated ice discharge from West Antarctica. In this case melting is not the primary cause of loss of ice mass. A more extreme impact comes from the effects of warming from warm seas on ice shelves like Larsen. As can be seen in Figure 9.7, the ice shelf, which is floating buoyantly on the water, can be undercut as warmer deep water melts the ice underneath it. When combined with surface effects, the sheet becomes unstable and breaks apart in the process of calving. Eventually the entire ice sheet disappears (as Larson A, B, and C have), as shown in Figure 9.8.

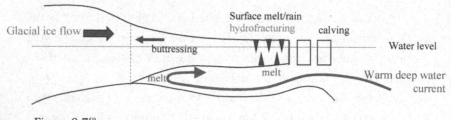

Figure 9.7[59]

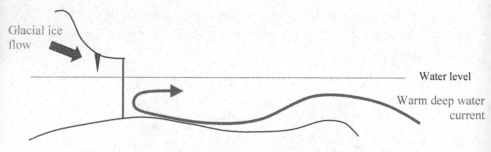

Figure 9.8

At this point, not only can the ocean current continue to undercut the remaining ice wall, but the buttressing provided by the floating ice shelf that has been holding the glacier back is removed. This allows the glacier to flow unfettered into the sea.

After the Larson B ice sheet disappeared, the glaciers feeding it sped up by up to a factor of eight, contributing an additional .07 mm/yr. to sea-level rise. A similar effect was observed when Larsen A ice sheet was lost.

Of greatest concern, however, is the Amundsen Coast Basin that contains the Pine Island, Thwaites, and Smith Glaciers, among others. These are not currently buttressed by large ice shelves and are accelerating and free-flowing at more than three times the speed of the average Antarctic glacier. Even though the glaciers that would feed ice shelves on the coast have been accelerating, the ice shelf hasn't been growing, indicating that they are probably melting from below due to warmer water. (As this book went to press, a *Proceedings of the National Academy of Sciences* study suggested even more rapid breakdown of both Thwaites and Pine Island shelves than previously thought.)[60]

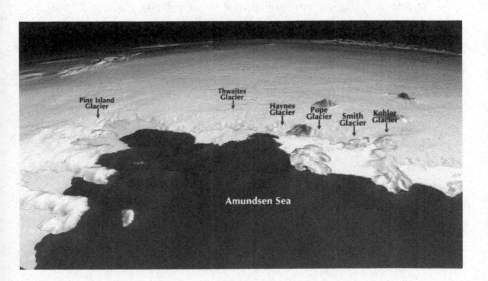

Currently these are contributing an ice-mass loss in excess of accumulation that accounts for between 0.14 mm/yr. and 0.24 mm/yr. Suppose just these three glaciers continue to accelerate at their current rate and double their current velocity by the end of this century, as models suggest, while the rest of the ice on the Peninsula continues to lose mass at a rate consistent with the current surface ice mass balance. Then by 2100, mass loss in the Peninsula will contribute at least 0.12 meters to global sea level rise.

That is significant, but it could be much worse—albeit depending on the accuracy of detailed model estimates of the glacier dynamics—and on a quicker time frame than many centuries or millennia. Two studies in 2014 suggested that these systems have already destabilized, and their disappearance within a century or two is inevitable. What makes this worrisome is that if released, the ice currently stored in these basin glaciers would produce 1.5 meters of sea level rise.

The situation could get even worse. A study in the *Proceedings of the National Academy* in 2015 modeled the glacier systems and fit them to their current rate of motion and melting. It concluded that even if the temperature is turned back down to its value in the 1970s and 1980s by some as-of-yet-unknown effect, not only does the destabilization of theses glaciers still continue, but the entire West Antarctic ice sheet becomes destabilized. Over centuries to millennia this would add an additional three meters to global sea levels.

We are now fully in the realm of disastrous "might be" Ghost of Christmas Future scenarios, which, while plausible, are not yet firmly established. So instead let's step back and tally up what we have ascertained is more or less guaranteed in the way of sea level rise by 2100. If one extrapolates from current heating and current changes in ice mass loss rates in Greenland and

Antarctica, and adds that to the already built-in thermal expansion of the ocean, as well as the impact of the melting of other glaciers and ice caps around the world, one arrives at:

Global Sea Level Rise Built-In by 2100:

Thermal expansion will contribute	≈0.23 ±.09 meters
Other glaciers and ice caps will contribute	≈0.37 ±.02 meters
With current acceleration of melting, Greenland will contribute	≈0.16 meters
With current acceleration of discharge, Antarctica will contribute	≈0.12 meters
Total	≈0.88 ±.12 meters

You get a sense of the extrapolation required to produce the above estimates for Greenland and Antarctic ice mass loss from Figure 9.9, which shows the relative measured contribution of each to ice mass loss and sea level rise for the period 1989–2018. The growing significance of Greenland ice melt is clear, and the downward curvature gives a sense of the current acceleration in the rate of mass loss.

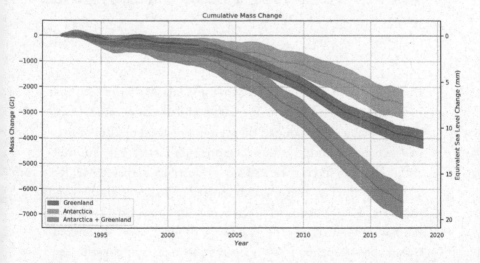

Figure 9.9

If instead the current accelerated ice loss in Greenland levels off, then it will contribute only 0.05 meters to global sea level rise by 2100. If at the same time, the West Antarctic Peninsula stabilizes so that the East Antarctic ice growth balances the West Antarctic ice loss, and Antarctica as a whole contributes nothing to net sea level rise this century, then the *minimum expected sea level rise*, already built in to the planet on the basis of the radiative forcing thus far, will be 0.65 ±0.12 m by 2100.

In other words, the minimum sea level rise we can expect this century is 0.5 meters, while the maximum we can expect, assuming no catastrophic collapse of glaciers in Antarctica or increased acceleration of melting in Greenland due to other climate feedback effects, is 1.0 meter. These two values bracket the expected range for the best-case scenario of the future *as it will be*.

Faced with measured sea level rise of a mere 3–4 cm since the turn of this century, it is easy for many people to view concerns about future sea level rise in the coming fifty to eighty years as premature. And for some, it may seem that even a 0.5–1.0-meter sea level rise is not such a big deal. But as I shall later elaborate, it is.

———

I have saved what may be the worst for last. We have seen that increased CO_2, leading to increased global temperatures, will result in sea level rise due to the well-understood physics of thermal expansion and also due to more subtle effects from differential changes in climate and ocean currents across the globe. In spite of the fact that the extrapolations to 2100 and beyond that I have described arise from a vast amount of data that has been accumulated over the past thirty to forty years, some skepticism about depending solely on climate models might still seem justified. Significantly larger sea-level changes in this century or millennium may still seem like the stuff of science fiction.

There is, however, one more set of compelling data that I have not yet shown. That involves paleoclimate measurements of the type I discussed for Greenland, but in this case ones that go back well beyond the direct CO_2 and temperature record and that also trace global sea levels over this time.

Using a host of markers, from coral reefs to isotopic abundances to coastal traces in stable geological structures, a clear correlation between CO_2 abundance, ice volume, and global sea levels emerges. If one goes back more than sixty-five million years ago to the age of dinosaurs, when atmospheric CO_2 was as high as 1250 ppm, sea levels were as much as 73 meters higher than they are now. By thirty-two million years ago, atmospheric CO_2 decreased to about 500 ppm—*comparable to what they are likely to be by the mid-twenty-first century*—and sea levels had subsided to merely 28 meters higher than at present. On the other hand, during the last glacial maximum just twenty-one thousand years ago, the sea level was as much as 130 meters lower than it is now.

Again, it may be tempting to dismiss extrapolations from millions of years back when direct measures of CO_2 and global temperatures are not available. However, the Red Sea is particularly sensitive to sea-level changes because of the narrow and shallow connection with the open ocean, so that lowering of the sea level decreases transport through the channel and increases the residence time of water in the Sea. Careful modeling of flow in this channel, combined with oxygen isotope measurements in sediment cores, then give estimates of sea level as a function of time. Estimates of sea-level low points during periods of maximum glaciation have been independently corroborated by other means, and the estimation method has been extended to allow continuous records of sea levels with careful attention to achieving time resolution on the order of a century where possible. For the past twenty thousand years Red Sea estimates of global sea levels can be compared to coral reef data taken from around the world and good agreement is obtained. Figure 9.10 is a recent compilation.

A reduction in sea level on the order of 120 meters during the period of maximum glaciation, mentioned earlier, can be clearly seen. Equally important is the fact that the sea level has varied by more than 10 meters within a few centuries, as can be seen during the period between fourteen and fifteen thousand years ago—compatible for example with a total loss of sea ice in Greenland within a millennium.

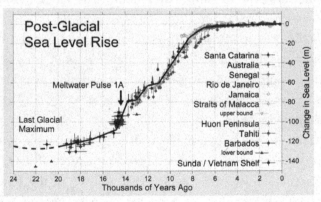

Figure 9.10[61]

For those who doubt significant sea-level change is possible, or who doubted the predicted relatively direct connection between CO_2 levels, global temperatures, and sea-level rise, I now finally turn to the figure that first shook me out of my own complacency in this regard.

Using Red Sea data, one can produce a curve of inferred sea level going back 440,000 years, which allows a comparison with ice core data. Figure 9.11 is from Vostok ice cores I showed earlier comparing measured CO_2 abundance with temperature.

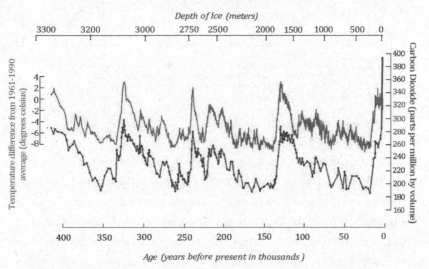

Vostok Ice Core Record:
Carbon Dioxide versus Temperature.

Figure 9.11[62]

Figure 9.12, by Jim Hansen, is of the inferred global sea level, based on Red Sea data.

It is true that correlation is not causation, but if one has a physical model, as we do, that predicts such a correlation, then observing one as clear as the one shown here between CO_2 abundance and sea levels is highly suggestive, to say the least.

Finally, I remind you that the current CO_2 atmospheric abundance of 415 ppm is already off the upper chart and the predicted CO_2 in 2100 will be even higher, almost a factor of two larger than any datapoint shown over the past four hundred thousand

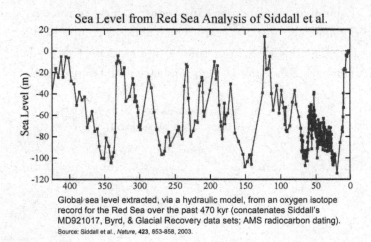

Global sea level extracted, via a hydraulic model, from an oxygen isotope record for the Red Sea over the past 470 kyr (concatenates Siddall's MD921017, Byrd, & Glacial Recovery data sets; AMS radiocarbon dating).
Source: Siddall et al., *Nature*, **423**, 853-858, 2003.

Figure 9.12[63]

years, during which sea levels varied by over 130 meters. While the temporal resolution of these plots is not sufficient to reliably predict what is likely to happen over the course of a single century, we can, on the basis of the historical record, assess that a likely sea-level change on the order of meters in a few centuries is not implausible, and thus decide with more confidence whether planning or acting now in the face of such a possibility is worth the risk.

CHAPTER 10
CLIMATE CHANGE TODAY

Climate is what you expect;
weather is what you get.

EDWARD LORENZ, quoting a proverb

EARLIER I WROTE THAT THERE IS A DIFFERENCE BETWEEN "weather" and "climate" because this distinction is relevant to how we interpret our daily lives. One anomalously cold day in Washington, DC, doesn't imply no global warming, and one ultra-hot week in the summer can't necessarily be blamed on it either.

It is worth considering, however, whether any sort of weather measurements might ever yield a statistically significant probe of climate change. If you measured the weather conditions everywhere around the world on a given day, or measured the weather in one location for a full season, could you discern definitive evidence that radiative forcing is causing climate change?

Recently a Swiss team of researchers claim to have demonstrated the first possibility. Using advanced algorithms and data for both temperature and humidity for many locations around the world each day since 2000, they claim that they can pinpoint data for any day that provides statistically significant evidence of global climate change.

Their specific claim is that by using temperature and humidity readings from around the world on a single day, they could work backward to extract a "fingerprint" of radiative forcing and hence warming. The fingerprint is determined by machine learning algorithms that explore the space of model predictions and compare these to the data obtained from around the globe.

This is a bold claim and the statistical techniques used by this group are far from intuitive. Their analysis, to my knowledge, hasn't yet been repeated by other groups. Nevertheless, their result is perhaps not as unexpected as it sounds on first hearing.

In a single location, weather can vary by tens of degrees from the mean temperature (as measured over some sufficiently long time span) on any day from year to year. This is why a snap freeze somewhere in April or a hot day in Washington in December doesn't provide evidence for or against global climate change. But if one compares temperatures from many different locations around the world on any day to mean temperature at each location that day, and then averages over the variations, one finds that large random daily variations in various single locations tend to cancel out from the full data set. Basically, it is an example of the statistics of large numbers. Just as individual humans often behave inexplicably when considered individually, when one averages over all of human population one may isolate statistically significant trends that yield insight into the human condition.

Shown in Figure 10.1 are their results from a simpler analysis than the full fingerprinting project that demonstrates the power of using a global analysis to explore even daily data.

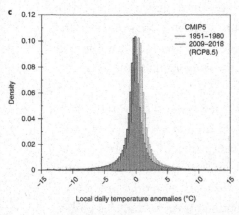
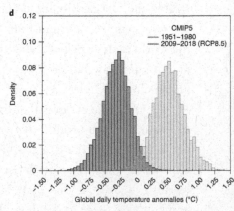

Figure 10.1[64]

The figure on the left is built up by recording the variation in the local daily temperature and comparing it to the mean local daily temperature for that day in that location. The distribution of such variations is then compiled for many different days. Then this distribution is combined with similar data sets from other locations around the globe to get good statistics for the average distribution of daily local temperature variations.

The second method first averages the temperature deviation on a single day over many different locations around the globe. Some will be positive and some negative, and the average variation will thus be smaller. One then plots the distribution of these averages for many different days.[1]

Two things jump out. First, the statistical distributions behave as expected for large data sets. They are roughly symmetric about the mean and distributed with more or less Gaussian shapes. Next, as expected, the distribution of average temperature variations on a single day for a single location has a much bigger spread than the distribution of the variations on that day when first averaged around the globe.

Because of the smaller spread of global daily temperature variations, the measured distributions for different eras can be more effectively compared. Here the global average daily temperature distributions during two different periods, 1951–1980 and 2009–2018, are plotted versus the mean daily temperature from 1979–2005. The fact that most of the range for the first period is negative implies that an average day was colder during the period 1951–1980 than it was from 1979–2005. The opposite is true for the average day during 2009–2018.

---

1   In the plots, the acronym CMIP5 in the legend stands for "Coupled Model Intercomparison Project Phase 5," which is a data archive used to analyze climate models and from which these data were extracted.

The difference of about 0.75°C between the peaks of the two distributions from these different eras suggests that even daily estimates, normally taken as evidence of weather and not climate, can, if suitably globally averaged, provide a signal of climate change.

If global daily temperatures can provide clear evidence that global warming has recently occurred for periods separated by decades of time, can one find some other *local* evidence that a global warming signal can be clearly separated from natural variability from one year to the next? The answer is yes, at least for some locations, provided one considers not daily temperatures but seasonal temperatures.

As the question suggests, the issue comes down to separating signal from noise. Shown in Figure 10.2 is a distinct oscillating signal. Below that, the same signal is displayed with significant noise overlaid on it. If you had just the second piece of data would you be able to infer the original pattern of oscillation?

Pure Oscillating Signal and Oscillating Signal with Noise

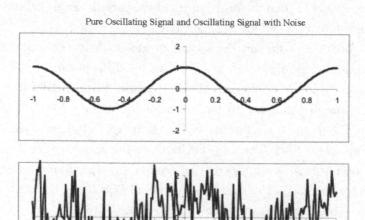

Figure 10.2[65]

The climate scientist Susan Solomon and colleagues have explored separating signal from noise by minimizing noise. They examined regions and seasons where natural year-to-year variability in temperature has been shown to be small and found that the warm season in equatorial regions seems to fit the bill. Figure 10.3 shows results displayed with different shades representing the difference in average temperatures for 1990–1999 vs. 1900–1999, for model predictions (left) vs. observations (right).

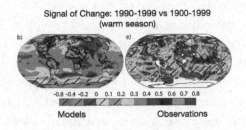

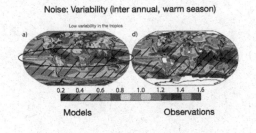

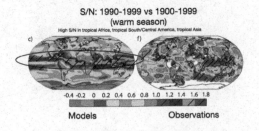

Figure 10.3[66]

In all cases, observations are noisier than the models, as one should expect, since the models don't contain all the messiness of the real world. But during the warm season in the tropics, circled in red in the left of the lower two sets, year-to-year variation is small in both the models and data. This is also not surprising. For anyone who has travelled in these regions, the weather report is almost always the same, something like eighty-eight degrees at 3:00 p.m. with a chance of rain, for example.

By comparing signal to noise in the model predictions, the significance of the signal in the tropics is clear but is visually less obvious for the actual data. However, if one plots the observed warm season temperature range each year versus the more-or-less-constant variability from year to year for a variety of locations around the world—from the 2.5°C average variability from summer to summer in Russia and ±2°C in the US to the ±1°C average variability in Central America and equatorial Africa—a clear signal emerges in the equatorial region. By 2020 the summer there is measurably different than it has been in the past (Figure 10.4).

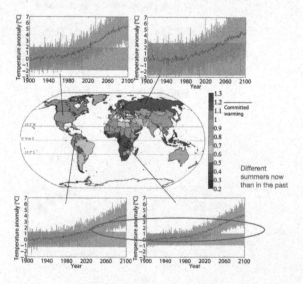

Figure 10.4[67]

The first regions that therefore are already experiencing clear effects of global warming on a year-to-year basis are the equatorial and tropical regions of the world. What makes this particularly unfortunate is that these include the countries that have been, on the whole, least involved in contributing to the cause (in the black box on the graph), where CO_2 emissions in tons per year per person in 2006 are shown in Figure 10.5.

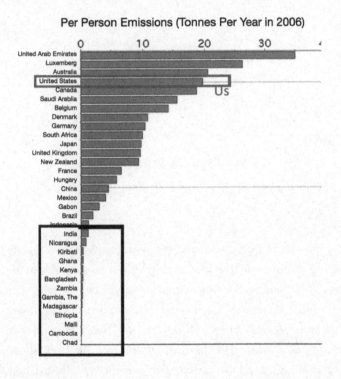

Figure 10.5[68]

These low-emission countries are also probably the least capable of mediating the effects of global warming. Unfortunately there seems to be a clear inverse relationship between the level of development in a country and signal-to-noise of global warming as Figure 10.6 conveys (data shown for 2009).

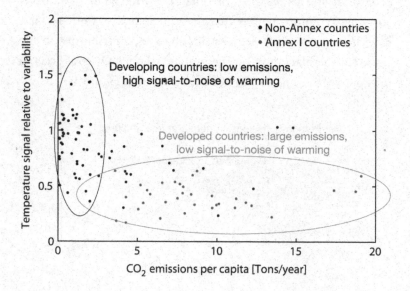

Figure 10.6[69]

The situation is likely to only get worse for those regions experiencing the immediately detectable effects of global warming today. By the end of the century, while it will be hotter everywhere on average, in the tropics and subtropics *the seasonal average summer temperature will be warmer than the warmest summer thus far almost 100 percent of the time.* It is in these regions that the greatest percentage of currently malnourished individuals live and where the population depends most heavily on agriculture, which will be most strongly affected by global warming.

When I was a boy, I worked at my parents' small gift shop in a local mall, often monitoring the movements of the kids who came in to buy penny candies. In the next aisle, above the other items for sale, hung a sign that read: *If you break it, it's yours!* If the climate is broken in any way, it will have been the developed world that has broken it, but it is the developing world that may suffer first. Ownership is clear, but who accepts ownership remains to be seen.

CHAPTER 11

WORST CASE: FROM FEEDBACK TO TIPPING POINTS

Nothing is written!

— T. E. Lawrence, from the film *Lawrence of Arabia*

Perhaps nothing strikes more terror into the heart of a numerical modeler than the words "nonlinear feedback." The reason numerical simulations are so useful is that they can handle many small timesteps, and during small time intervals generally all changes are linear. Therefore even complicated behavior over the long run can be successfully simulated with relatively simple mathematical operations at each step.

The problem with nonlinear feedback mechanisms is that an initial small change can produce an effect that feeds back on the changing variable, producing yet a larger effect. Pretty soon untreatably large effects can grow exponentially fast. When this happens, predictive capabilities are rapidly lost. So goes the famous example of chaos, where a butterfly flaps its wings in Kansas, ultimately causing a tornado to rip through Nebraska.

But nature doesn't care about the happiness of modelers or the utility of their modeling techniques. There are undoubtedly feedback mechanisms that affect climate that can be difficult to model because of the sudden changes they may bring about.

Here is one possible example. Heating causes ice to melt in Arctic waters. But the melting ice exposes the darker water underneath, which reflects less sunlight and absorbs more, causing more heating that melts more ice, and so on.

Here is another: ice melts in Greenland and reduces the height of the central ice sheet, but the temperature at lower altitudes is higher, so as the top of the ice sheet lowers in elevation, it heats up yet more and melts more quickly.

Here is one ice possibility I previously described: warm water causes floating ice sheets to calve and eventually break apart, but these buttress large glaciers that now are unrestrained, and quickly flow into the sea, destabilizing the entire ice sheet behind them.

This book, and all my others for that matter, aim to demonstrate how an explanation of fundamental physics arguments can give interested laypeople the tools to: (a) address questions they may have about the world and their place within it, (b) assess claims about the world they may read about, and (c) make decisions about their own actions and about public policy questions. With this in mind, I have concentrated on the general physical principles and tools that give some basic perspective on how a predictive understanding of the causes and various effects of radiative forcing on climate can be obtained and how reliable it might be.

There are some important potential effects that are, however, difficult to connect directly to underlying physics or to include directly in models. This does not mean we should ignore them. One of the most powerful and important aspects of science is its acknowledgment of uncertainty. The fact that all physical statements have uncertainties associated with them is not a failing but a virtue. It doesn't mean that we cannot predict results reliably nor that we don't understand physical systems. It means we must work to quantitatively understand and explicitly state just how reliable a prediction is likely to be. Thinking deeply about the quantitative uncertainties in estimates or measurements is a central part of making the estimates or measurements.

That the basic predictions of climate science come from well-understood physical principles should make it clear that much of climate science is not some invisible voodoo practice or something that requires supercomputers to assess. Because of this, we all should be able to appreciate and assess the causes, effects, and risks that arise from these predictions and that we face as a result of human industrial activity.

I would be remiss if I didn't follow up on the preceding discussions of these predictions by at least briefly pointing out some of the "known unknowns"—those less predictable consequences that are based on the interplay between numerous climate factors that feed back on each other. Recent work highlighted in a *Nature* opinion piece in November 2019 describes a number of these. We may not be able to state with authority when or if any of them will become significant, but in any risk analysis we would be negligent not to at least factor them in at some level into our decision-making.

⋙⋘

The effects of greatest concern for climate modelers, and the rest of us, are what have become known as "tipping points"—those drastic changes, over various time frames, that once set in motion are likely to be irrevocable. I have already discussed one such tipping point, the West Antarctic Ice Sheet. Once destabilization of the glaciers in the Pine Island Bay of the Amundsen Sea begins in earnest, then at least one study suggests that even if temperatures return to pre-2000 levels, the entire West Antarctic ice sheet may be destabilized. This would add perhaps three meters to sea level rise over the course of a millennium or two. Adding some credibility to this possibility, there is paleoclimate evidence that such widespread collapse of the ice sheet has occurred more than once in the past. Once floating ice shelves that buttress glaciers are lost, there is nothing to stop the irresistible pull of gravity bringing glaciers down into the sea.

Other feedback effects I have mentioned that happen to be related to ice could lead to a complete collapse of the Greenland Ice Sheet on a timescale shorter than that predicted due to temperature rise and ice mass melting. That prediction suggests a temperature rise of 6–7°C would be necessary for irrevocable collapse to occur, and even then over a timescale of multiple millennia. However, there is sufficient uncertainty so that a much smaller temperature rise could lead to collapse. Recall that one unanticipated factor that played a role in the 2019 severe melting summer was the large high-pressure zone that centered on Greenland, probably due to ocean current changes, leading to more sunshine and melting and less precipitation. Another effect mentioned earlier is the nonlinear feedback that occurs as ice sheets melt: their elevation decreases, putting the surface in contact with warmer air, which causes the sheet to melt faster.

Effects such as these, combined with the increased relative warming in the Arctic water due to changing albedo and increased absorption of light as sea ice melts, have led some researchers to suggest that the Greenland ice sheet threshold, beyond which it is doomed to melt, could be as low as 1.5–2°C. This number is ominous because we have already basically committed the Earth to such a temperature rise before mid-century, given the CO_2 already in the atmosphere. Of course even if Greenland ice-sheet melting is assured, with its consequent seven-meter sea level rise, the time period over which this could happen may depend on the future temperature profile of Earth. Thus actions taken in this century could perhaps delay the inevitable beyond this millennium.

Another area where irreversible sudden shifts may be possible involves large-scale ecosystems, an area I have not focused on in this book, not because it isn't important, but because the issues here are more biological than physical. One particular biological impact that has received some international attention is the mass bleaching of large coral reefs, with the Australian Great Barrier Reef being the most famous. The severe bleaching of much of this reef has already been documented with photographs that have appeared in newspapers around the word. Marine biologists project that if the temperature rises by a further 2°C, 99 percent of the world's tropical coral reefs could be lost. Multiple factors could come in to play here beyond simple warming, although warming dominates. One in particular is worth reviewing here because, like sea level rise, it involves straightforward physics and chemistry directly associated with increased CO_2 concentration in the atmosphere.

I have described how an equilibrium is built up between CO_2 in the atmosphere and the formation of carbonic acid in the oceans as CO_2 dissolves. One of the important observations that can be used to argue against those who claim that increasing CO_2 is having little global effects (and by consequence, that global warming is related to something else, like rapid changes in solar luminosity, which, by the way, we can measure and which do not exist) is that the world's oceans are already responding to a different kind of forcing from increased CO_2 in the atmosphere. We are apparently entering a new phase, some call it Oceans 2.0, due to both ocean warming and ocean acidification.

As early as 2002, ocean measurements made it clear that the mean pH level of the world's oceans had been decreasing at a regular rate for over a decade. Measurements have continued since then, and the chart in Figure 11.1 was prepared by the US National Oceanographic and Atmospheric Administration (NOAA) in 2018.

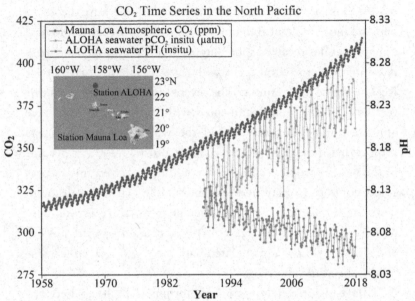

Data: Mauna Loa (ftp://aftp.cmdl.noaa.gov/products/trends/co2/co2_mm_mlo.txt) ALOHA (http://hahana.soest.hawaii.edu/hot/products/HOT_surface_CO2.txt)
Ref: J.E. Dore et al, 2009. Physical and biogeochemical modulation of ocean acidification in the central North Pacific. *Proc Natl Acad Sci USA* **106**:12235-12240.

Figure 11.1[70]

Shown in the figure is the CO_2 abundance from the Keeling curve in Mauna Loa we have already seen, and alongside it are measurements in the ocean off Hawaii of the amount of CO_2 dissolved in seawater and the seawater pH. As would be predicted in any high school chemistry text, as the CO_2 atmospheric abundance increases, the fraction of CO_2 dissolved in water in contact with the atmosphere will increase. As that happens, the amount of carbonic acid, H_2CO_3, will increase, which in turn will lower the pH.

Decreasing pH in the oceans may have many complex biological effects, but one that involves simple chemistry is this: as
oceans become more acidic, the increase of positive hydrogen
ions will allow their increased binding with pre-existing negatively charged carbonate (CO_3) ions. But the carbonate ions are
what both coral and shellfish use to build their calcium carbonate for their exoskeletal structures. With fewer carbonate ions
free for the taking, corals will suffer as will shellfish. Indeed,
a study of coastal waters off the west coast of the US in 2011
surprisingly found that 53 percent of tiny pteropods, very small
marine snails, already had severely dissolved shells. NOAA has
done experiments on these small snails by placing them in water
prepared with the ocean acidity predicted in 2100 for forty-five
days. Figure 11.2 is a photograph of the time series.

Figure 11.2[71]

As dismal as these photographs are, to be fair one must be
careful before inferring increased ocean acidity will immediately
bring disaster. The consequences of a reduction in oceanic biodi-
versity involving a possibly reduced fraction of pteropods and a

closely associated shelled zooplankton called foraminifera could be severe, as these small animals are an early step in the food supply chain for almost all fish in the ocean. But if they disappear, will that simply create an evolutionary niche for other non-shelled zooplankton to adapt to take up the same role?

Also, when these shelled animals die, they sink to the seafloor and carry their calcium carbonate with them. This gets deposited as rock, sequestered, and eventually subducted into the mantle, or as with the White Cliffs of Dover, raised above sea level by plate tectonic movements. How might their possible disappearance impact the amount of CO_2 taken up by the oceans? Naively, one might think it would reduce this intake, which would lead to increased CO_2 in the atmosphere—another feedback loop. But the carbonate chemistry of ocean water involves more than just calcium carbonate, so inferring a direct feedback on atmospheric CO_2 in this way is too simplistic. A variety of competing effects will determine the ultimate impact, if any.

On larger scales, to what extent various large fish populations, a staple for feeding much of humanity, will survive a combination of increased temperature and increased acidity is currently not known. But clearly a collapse of any major fish species would reverberate throughout the planet.

Feedback on land may also produce climate change tipping points, including the impact of warming on North American boreal forests, where large-scale insect infestations and increases in forest fires have caused dieback. This turns forests from carbon sinks to carbon sources, not just directly through fires but also through decay of dead trees and plants.

A potentially more severe feedback loop may occur in the Arctic, where warming can melt the permafrost, which contains huge stores of CO_2 and methane—perhaps as much carbon could be released in this way from the permafrost of Siberia, full of organic material, as currently exists in the atmosphere. And methane is an even more potent greenhouse gas than CO_2. Permafrost is already being observed to thaw, and at what temperature the thawing may become irreversible remains to be seen.

Increased greenhouse forcing is not the only impact of human industrial activity that may produce a nonlinear climate or biological tipping point. Deforestation of the Amazon has been significant, with about a 17 percent loss of Amazon rainforest to date. But the humidity in the region is strongly impacted by the existence of the rainforest itself. At some tipping point, the forest stops increasing the humidity and rainfall so it is no longer self-sustaining, and it will degrade into a drier savannah. This would release huge amounts of CO_2 currently stored in the rainforest trees and vegetation into the atmosphere. One estimate, by Thomas Lovejoy, is that the Amazon tipping point could happen when 20–25 percent of the rainforest has been cut down. Given the current 17 percent level of deforestation, this could happen before midcentury. According to at least one economist, the combination of deforestation, large-scale forest fires, and global warming could accelerate the rate of destruction, meaning the tipping point could be reached in this decade.

As worrisome as each of these possible tipping points—some leading to short-term disasters and others, like the potentially irreversible melting of Greenland, taking centuries or millennia—may separately be, of even greater potential concern is a possible feedback between them. There is increasing evidence that such feedback may be common. A 2018 study of thirty possible systems at risk found that in 45 percent of cases, a tipping point in one area had a domino effect, increasing the risk of exceeding a tipping point in another.

As an example, here is one scenario that the authors of the 2019 *Nature* opinion piece argued is already currently underway:

> "Arctic sea-ice loss is amplifying regional warming, and Arctic warming and Greenland melting are driving an influx of fresh water into the North Atlantic. This could have contributed to a 15% slowdown since the mid-twentieth century of the Atlantic Meridional Overturning Circulation (AMOC), a key part of global heat and salt transport by the ocean. Rapid melting of the Greenland ice sheet and further slowdown of the AMOC could destabilize the West African monsoon, triggering drought in Africa's Sahel region. A slowdown in the AMOC could also dry the Amazon, disrupt the East Asian monsoon and cause heat to build up in the Southern Ocean, which could accelerate Antarctic ice loss."

Lest any or all of these seeming doomsday scenarios make you want to crawl under a desk (or burn this book, except for your worries about that contribution to the greenhouse effect), I want to stress that all of these examples are of a future that *might be*. None can be as simply or directly modeled as the impacts I have focused on in the rest of this book, and none of them are guaranteed. But it is worthwhile recognizing that any one of them could have a short-term impact that could be more devastating than the slow and steady rise in global temperature might otherwise suggest.

It will be up to decision-makers to determine if we are willing to empirically test these ideas in the coming decades. Or whether the risks, which could be small, when multiplied by the possible devastation, which in some cases could be immense, combine together to convince politicians and the public that we need to act globally now.

That is, I hasten to add, if it is not already too late.

CHAPTER 12
BACK TO THE MEKONG

It's a funny thing coming home. Nothing changes.
Everything looks the same, feels the same,
even smells the same.
You realize what's changed, is you.

ERIC ROTH

AS WE RETURN BACK TO THE MEKONG, RECALL THE FLOOD OF
feelings I earlier described that overcame me as I looked around
from the bow of that riverboat as it churned along the Mekong.
I had spent some months before boarding the vessel on an intellectual journey into the world of climate change science. Without
context, the fact of climate change and its possible impacts can
easily seem far removed, perhaps because at the same time they
can also seem so overwhelming. The result can be stultifying.

But as I experienced the Mekong that day, I looked around at
my surroundings with new eyes. My abstract intellectual journey
had become much more personal. The beautiful landscape I saw
all around me was in urgent peril, making the possible impact of
climate change both tangible and immediate.

I began this book with John Keay's description of that mighty
intertwined system of river and humanity, along with the para-
dox of its existence. It is worth repeating here:

This diurnal mother-of-a-tide ought, of course, to spell disaster to the Delta. A salty inundation, albeit only once a day, would soon sour the world's most productive rice-bowl and turn the green dazzle of paddy into maudlin thickets of mangroves like those along the Donnai below Saigon. What prevents such a disaster is the power of the mighty Mekong. The inrushing tide meets the outrushing river, and in the best traditions of ecological equilibrium they compromise. The river rises, its progress barred by the tide. The backing-up of the river by a big 'diurnal' is measurable as far upstream as Phnom Penh and beyond. But there and throughout the three to four hundred kilometers down to the sea, salination is barely detectable… The river thus defends the Delta from its deadliest foe since the rising waters are overwhelming its own, not the China Sea's.

Sadly, this is a battle the Mekong is destined to lose. Its demise will be swift, brought on by forces that originate from far away.

The story I will now relay involves two completely independent scientific studies, each done with tools as different as one can imagine. They were reported within weeks of each other in 2019, both leading to startling conclusions with implications not just for the Mekong but for the rest of the planet.

The first study involves my favorite kind of science, at least when I pursue it myself: feet on the ground and low tech.

Dutch geographer Philip Minderhoud, a postdoctoral researcher based at Utrecht University, noticed something strange while visiting Vietnam a number of years ago to investigate land subsidence—the lowering of ground level due to natural and human-induced processes. Comparing what he saw on the ground to satellite maps, he realized something was wrong:

"I saw, for example, roads that were only slightly higher than the water that was directly connected with the sea, whereas according to the maps, they should have been located higher." He also saw strange patterns of contour lines on the maps, linear stripes that clearly weren't real traversing riverbeds, for example. Clearly the satellite maps themselves were suspect.

A technique called LIDAR, which stands for Light Detection and Ranging, uses the reflection of pulsed lasers from the air to measure ground elevation, and it is accurate to thirty centimeters or better. While there are some LIDAR systems in space which have been used to measure elevations of the Earth's polar regions, generally LIDAR has been used by planes and is expensive. Traditionally, therefore, large-scale global land elevation maps are done using other kinds of satellites which create Digital Elevation Models (DEMs), most often using radar techniques. The resolution of these DEMs typically range between one and thirty meters.

For example, in February of 2000 the space shuttle *Endeavour* engaged in a mission called the Shuttle Radar Topography Mission (SRTM), a global survey sponsored by the Department of Defense. Its data is publicly available and has been widely used by various organizations, including the World Bank, to decide where to allocate flood control resources.

However, there are problems using this data to model low-lying deltas. In particular, there is a difference between absolute elevation and relative elevation. To turn one into the other, one must have calibration points—predetermined zero points of elevation, which themselves can vary around the Earth due to local effects. Earth isn't a perfect sphere, and the gravity of land masses will impact on local sea levels as well. What one needs is a good local calibration. Without this, satellite elevation readings can be inaccurate, especially for low-lying areas. And things become even less accurate if calibration points are based on land masses that may themselves have been subsiding.

Because the Mekong Delta is built up from soft river sediments, when it compacts it can subside. In addition, upstream dams can block the flow of new sediments to replenish those lost by erosion. These effects can be exacerbated when large urban areas pump up groundwater from below and where urban infrastructures interfere with the natural seepage of rainwater into the Earth. In Mexico City, one of the world's largest cities, built on what was originally a large underground aquifer now being drained to provide fresh water for over ten million people, settling is occurring at a rate of up to twenty centimeters per year. It is quite noticeable in the sagging of building foundations as you walk down various streets. In the Mekong, subsidence in some areas has been measured at rates approaching five centimeters per year. It was in this context that Minderhoud began his studies.

Satellite and space shuttle data had put the elevation of the delta at 2.6–3.3 meters, uncomfortable but tolerable given sea level rise estimates for this century. Apparently, the Vietnamese government had already known that the elevation maps used by the West did not conform to their own elevation readings, but the detailed elevation data was a military secret.

It was here that Minderhoud exploited the human side of science. He built trust with the Vietnamese authorities, pointed out how his research might contribute to their own goals, and offered to combine his research program with their efforts. In the end, he was given the data for twenty thousand separate elevation points throughout the Delta. In return, he significantly improved the calibration of these points. The Vietnamese had used a point near Hanoi, two thousand kilometers to the north. Instead, Minderhoud's group used a local point for calibration at an island location, Hon Dau.

The result of all of this on-the-ground effort was a new topographical Digital Elevation Model tailored to the region, and the change was staggering. It turns out the new mean elevation of the delta is about 0.8 meters above sea level, about 2 meters below the older satellite estimate.

Figure 12.1, from Minderhoud's paper in *Nature*, gives a sense of the difference between the old satellite-based estimates of elevation (labeled STRM), a more recent model based on that data (MERIT), and the new more accurate topological model (Topo). Shown by black lines are two cross-sections of elevations across the country. The earlier DEMs overestimate the elevation at every single location. Minderhoud demonstrated that reasonable agreement between MERIT and Topo could be obtained by systematically lowering the MERIT elevation at every point by 2.5 meters.

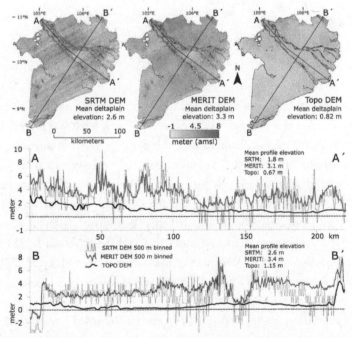

Figure 12.1[72]

Figure 12.2 is an aerial map obtained using Minderhoud's full data set, with different shadings displaying land elevation across the entire delta region.

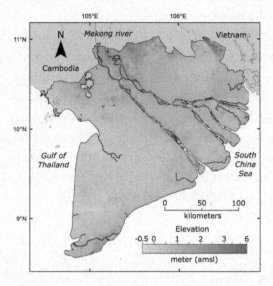

Figure 12.2[73]

The implications of this for the region are staggering. As we have seen, sea level rise is guaranteed to be at least 0.5 meters by 2100. But worse still, including the average current subsidence rate of the land of ≈1.0 cm per year, the relative sea level rise compared to the land elevation in this region will be at least 0.8–1.0 meters by century's end. Figure 12.3 is a plot showing how much of South Vietnam will then be below sea level.

The relevant statistics, assuming 1.0 m relative sea level rise are as follows: 75 percent of the total area of the Mekong Delta plain, comprising most of South Vietnam, will be below sea level. The number of people who currently live in this area is about 12.3 million people, about 70 percent of the total population of the region.

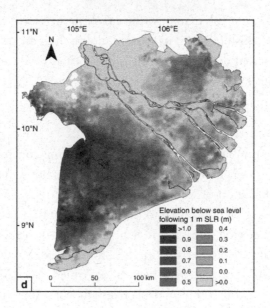

Figure 12.3[74]

The situation is actually even worse. The Delta currently provides a food store for a far larger population, through freshwater fishing and rice production. Both are vulnerable to collapse if the area is inundated with seawater. And finally, to exacerbate the situation, there is the incessant removal of sand from the Mekong itself. As I noted earlier, this has reduced the Mekong riverbed elevation by 1.4 meters since 2008.

Climate change, combined with other impacts of human activity in the region, does represent a Perfect Storm closing in on the Mekong Delta. While the daily battle between the flow of one of the mightiest rivers in the world and the incoming daily tide of the China Sea goes on each day, the war has already been lost.

The richest rice-producing area in the world could be submerged under salty water by the end of this century. A river with

159

more freshwater fish than any other river in the world will likely become brackish far from its current mouth. And an agricultural population that depends on both will lose their food source, and much of the land on which they currently reside.

It may seem premature to draw these damning conclusions on the basis of a single study, however impressive. But within weeks of the publication of this work, a completely independent and broader research program published a paper in the same *Nature* journal describing the use of new neural networks to recalibrate and globally reduce errors in the Shuttle Radar Topography Mission DEM, producing a new model called CoastalDEM. In the region in which this analysis overlaps with the Dutch team analysis, the agreement is remarkable. CoastalDEM concludes a population between nineteen and thirty million in Vietnam currently reside on land that will be below the future high tide line as early as 2050. Figure 12.4 is a pictorial representation based on their new data of the land (shaded) that will be below high tide and annual floods in 2050 in South Vietnam—essentially the whole country below and including the currently bustling megalopolis of Ho Chi Minh City (assuming no new man-made coastal flood defenses implemented).

While many factors present in the Mekong Delta make it particularly sensitive to the impact of rising sea level, the Mekong represents, in microcosm, the threat facing much of the world's population from sea level rise. The CoastalDEM study explored the global impact of their new elevation data. Here is their summary abstract (bolded text is mine):

"Here we show – employing CoastalDEM—that **190M people (150–250M, 90% CI) currently occupy global**

Figure 12.4[75]

land below projected high tide lines for 2100 under low carbon emissions, up from 110 M today, for a median increase of 80 M. These figures triple SRTM-based values. **Under high emissions, CoastalDEM indicates up to 630M people live on land below projected annual flood levels for 2100,** and up to 340M for mid-century, versus roughly 250M at present. We estimate one billion people now occupy land less than 10m above current high tide lines, including 230M below 1m."

These are not the numbers of some remote future. They describe a future we are now more or less committed to in this century. It is hard to know how to adequately express the true human toll associated with staggering figures like this.

I have focused here on the Mekong Delta, not only because it will be hit so severely, but also because I cannot help but reflect on the friendly, generous, and vivacious citizens of Cambodia and Vietnam who I had the pleasure to meet and interact with. These people had suffered in the past century from the very worst atrocities of war and genocide, and yet emerged from this maelstrom without bitterness or a sense of victimization. Rather, they retain an optimism that the future can be better than the past, and they willingly shared with me the pleasure of our common humanity.

Cambodia rebuilt after a disaster largely of its own making in which perhaps a third of its citizens were massacred. Vietnam suffered a loss of millions of people in a war that was prompted from abroad to emerge as a thriving economic powerhouse today.

Both countries now face a threat far greater than they have ever faced, and one largely not of their own making.

The question of how these countries and their cultures might survive through to the other side of this century will reflect, in microcosm, how humanity in general will deal with a global phenomenon that is unprecedented in modern human history and about which we have had fair warning.

Context is everything. One meter does not seem like a huge elevation, unless perhaps someone is holding a bowling ball at this height above your feet. Or unless it is all that stands between you and the loss of your home, your livelihood, and in some cases much of your country by the middle or end of this century.

Vietnam is not alone. Without intervention, much of Bangladesh will disappear at high tide, as will large swaths of land in India, Indonesia, Malaysia, Papua New Guinea, and China in Asia; the Netherlands, Northern Germany, France, Spain, England, and Northern Italy in Europe; southern Iraq and northern Egypt in the Middle East; Mozambique in Africa; Argentina and Brazil in South America; Mexico in Central America; and Florida, Louisiana, Massachusetts, and North Carolina in the USA. Beyond this, almost every major coastal metropolitan area in the world will be partially submerged at some time during the year by 2100, including London, Tokyo, Miami, New York, Boston, and Shanghai, to name just a few.

The policies we adopt in the face of the information I have presented here—from the fundamental and well-tested predictions of radiative forcing to the more speculative possibilities of tipping points, and from the long-term global risks we face to the more immediate devastation facing places like the Mekong— are not for me to address here. They will be determined by the citizens of the world and the governments they elect or under which struggle.

I hope for the sake of Cambodians and the Vietnamese, and also for hundreds of millions of Bangladeshis, Indians, Japanese, Chinese, Africans, Polynesians, Indonesians, Middle Easterners, North and South Americans, Europeans, and all the rest, that technological ingenuity combined with rational action, tempered by simple grace and empathy, might supersede our longstanding human traditions of xenophobia, greed, and violence in the face of the national and international challenges we now face.

But hope and expectation are two different things.

EPILOGUE

FORTUNE FAVORS THE PREPARED MIND

If not now, when? If not you, who?
Paraphrased from Hillel the Elder

THIS BOOK ISN'T PRIMARILY A CALL TO ACTION, BUT NEITHER IS it a requiem.

As my friend Richard Dawkins often stresses, humans have a capacity for planning the future that is probably unique among all lifeforms on Earth. Evolution, while often anthropomorphized, has no plan by which it proceeds. Nature is sublimely indifferent to whether we survive as a species, or even whether our planet continues to exist.

But we humans, with our potentially unique capacity for self-awareness and rational thinking, can plan and strategize for the future. Moreover, we can develop scientific tools that allow us to predict the outcome of our actions as well as the outcomes of things we have no control over. And we can develop technologies that change the present and the future, that give us unprecedented control over our environment.

These remarkable gifts have brought us to the present moment, which some might view as a precipice and others view as a mountaintop, from which we can see farther than we have ever seen before. In the process, they have created a global civilization and a global economy we are still learning how to deal with as social beings.

I am writing this during the first global pandemic in which the interconnectedness of humanity has been manifest, and the impact of acting quickly in the face of evidence has never been clearer. I can't say that the global report card in this regard has yet been very impressive. But I also know that most of us feel certain that this, too, shall pass. We may have different levels of confidence on our political leaders, but we have confidence that our scientists and doctors can recommend courses of action that make us safer, and they will ultimately develop treatments and vaccines that can ameliorate the levels of illness and death that the current virus might otherwise cause.

The discomfort of our current regimen of enforced isolation and social distancing are real, but we expect they will be temporary. Life after the pandemic will undoubtedly be changed, but will it change more than superficially?

Imagine instead a global change far more pervasive and potentially far more long-lasting. Imagine not just your immediate livelihood under threat, but your home and the homes of your family and friends. Imagine the long-term livelihood of everyone you know potentially destroyed, with no safety net to fall back on, with no place to retreat.

This is the near-term future that seems likely for some significant fraction of the world's population, independent of the vagaries and vicissitudes of the longer-term impacts of climate change.

We need to face that future with open eyes if we are to plan for how to overcome it, or at the very least, ameliorate it. The only way to do that is to understand where we have come from and where we are heading, as well as what the impacts of our action, or inaction, might be.

I have many friends who are climate scientists, geophysicists, and oceanographers, and I am struck by their different emotional responses to the same reality, ranging across the gamut from stoicism to public protestation. But they share a fundamental optimism that rational action based on empirical evidence and the well-tested scientific models is the best hope we have to address the challenges we face in the coming century.

Among these individuals, perhaps the most optimistic, is my friend Daniel Schrag, who leads the University Center for the Environment at Harvard University and who was on President Obama's Council of Advisors on Science and Technology. We spent almost two hours on a recent podcast of mine discussing climate change, and his view is that technology provides tremendous opportunities for economic and social well-being in the face of climate change. These include transitions to more sustainably distributed energy resources, agricultural transitions that can feed more with less, and ultimately, if necessary, proposals to mediate radiative forcing to slow the temperature rise or even cool the planet—including geo-engineering options that involve adding things like aerosols to the atmosphere or, less realistically perhaps, global capture of carbon from the atmosphere followed by carbon sequestration.

I am not sure I am as optimistic, but neither do I see us traveling down an inexorable road to oblivion. The future is charging at us like a freight train, but it is doing so on tracks we have built. We may have time to divert the train, or perhaps build a bridge so it safely bypasses us. We will never know unless we try. I wish I had some silver bullet, but all I can offer is what I have tried to do here, to provide some perspective on the realities we face and how we can understand them.

I remember once appearing on television in Australia with some leaders of the Australian government and using a possible climate change consequence to impact their thinking as they prepared to walk back their country's climate change policy. If the Indonesian boat refugees so concerned them now, I argued that they should consider what it would be like to have potentially millions of climate refugees from countries for which Australia might be the nearest safe haven. My argument had no effect.

Scientists like me, not elected by the public, should not be making public policy. But we do have an obligation to use the results of our work to help inform public policy and, where possible, create technologies to aid in its enactment.

Technology could, in principle, ameliorate some of the inevitable impacts of climate change. But the greatest challenges we face are likely to be political and economic rather than technical. Developed nations may be able to protect areas that may lie below future high tide levels—as the Dutch did in the 20th century—at an increasingly arduous cost.   But what about less developed countries? Will assistance be provided to avoid mass migration? Improved agricultural technology could help keep family farms afloat in areas that will be impacted by climate change, but the hardest hit regions may also be the poorest, with least access to new technologies.

Understanding the science of climate change and its likely impact is the first step. Informed and willing policymakers will then need to weigh the safety, sociopolitical, and economic risks associated with future amelioration challenges versus the challenge of enacting potential preventative economic and technical measures now.

I have had cause numerous times over the past months to think back to the example of Louis Pasteur, the French biologist and chemist whose groundbreaking work contributed to the principles of vaccination, as well as the technique of pasteurization, named after him. His breakthroughs didn't end disease, and they didn't force government officials to listen to reason. The current nonsensical anti-vaxxers who are rabidly fighting against reality and the social contract, testify to the limitations of rationality among the human species.

Yet Louis Pasteur anticipated that when he said fortune *favors* the prepared mind. It makes no guarantees. It just offers better odds. I'll take that any day.

FURTHER RESOURCES

THERE ARE MANY PUBLICLY AVAILABLE RESOURCES RELATED TO material I have presented in this book. Here is a partial list, along with references to several of the specific scientific articles I have highlighted in the text. This list is not exhaustive but interested readers can turn to these sources for additional references.

Online resources

Our World in Data https://ourworldindata.org
This is a wonderful public resource of graphical data on a wide variety of topics. Moreover, all of the marvelous graphical presentations are freely available for use by others. One of the two best resources for graphical material on global climate change I have found.

Global Carbon Project https://www.globalcarbonproject.org
The Global Carbon Project provides a comprehensive set of resources related to climate change, from beautifully produced graphics to links to important ongoing studies. They produce the *Global Carbon Atlas*, which is a platform to explore and visualize the most up-to-date data on carbon fluxes resulting from human activities and natural processes.

National Snow and Ice Data Center, https://nsidc.org
A definitive source of data on the Arctic, Antarctic, and historical data on snow, glaciers, ice sheets, sea ice, ice shelves, soil, and frozen ground.

Arctic Report Card https://arctic.noaa.gov/Report-Card
This is an online peer-review resource issued annually since 2006 for reliable and concise environmental information on the current state of Arctic environments relative to historical records.

National Oceanographic and Atmospheric Administration https://www.noaa.gov
This is a comprehensive government resource on all things oceanographic and atmospheric. Besides the arctic report card, there are numerous articles with data and information on climate and climate change. The specific resources on climate are located in separate NOAA links below:

Climate.gov https://www.climate.gov
Here one can find charts on global average temperatures, CO_2 levels, snow cover, sea level data, sea ice, and ocean heat. 2020 Global Temperature data quoted from Jan 16, 2020 article on this site by Rebecca Lindsey and LuAnn Dahlman.

National Centers for Environmental Information. https://www.ncdc.noaa.gov
Another NOAA site, this has information and charts on historical climate data, including ice cores going back eight hundred thousand years.

Environmental Protection Agency https://www.epa.gov/climate-indicators
Useful historical data and discussions of radiative forcing, greenhouse gas concentrations, weather, and climate.

FURTHER RESOURCES

National Weather service. https://www.weather.gov/jetstream/energy
Here one can find information and graphs related to atmospheric issues relevant to both weather and climate, including the energy balance between the Earth, the atmosphere, and space.

NASA Earth Observatory https://Earthobservatory.nasa.gov/features/EnergyBalance
These web pages, while no longer being updated, provide good images of data and discussions of climate issues.

Scripps Institution of Oceanography https://scripps.ucsd.edu/programs/keelingcurve/
The Scripps Institution is where Charles Keeling worked continuously from 1956–2005, and they maintain the Mauna Loa measurement facility reporting daily on CO_2 levels. This site has information on current measurements and historical data, as well as much useful ancillary information.

National Academies Press https://www.nap.edu/catalog/12877/
This is a link to the comprehensive National Academies study on Climate in 2011, entitled Climate Stabilization Targets: Emissions, Concentrations, and Impacts over Decades to Millennia.

US ITASE http://www2.umaine.edu/USITASE/teachers/links.html
The International Trans Antarctic Scientific Expedition contains links to publications, data, and other resources for finding out about Antarctic research.

AntarcticGlaciers.org http://www.antarcticglaciers.org/glaciers-and-climate/ice-cores/ice-core-basics/
This site has valuable information and figures on ice cores, Antarctic mass balance, and historical climate information.

Climate Home News https://www.Climatechangenews.com
An online clearing house for up-to-date information relevant to climate studies and more.

Environmentcounts.org https://environmentcounts.org
This website provides an independent source of information on the environment, including figures for the 2017 compilation of ice core and temperature data for the past 420,000 years.

Realclimate.org http://www.realclimate.org
This website contains a series of informative articles on climate and climate science. Of particular relevance to material in this book are discussions in 2007 by Spencer Weart and R.T. Pierrehumbert on radiative forcing.

American Chemical Society: https://www.acs.org/content/acs/en/climatescience.html
This site provides a very useful Climate Science Toolkit for understanding various aspects of climate and climate change, radiative forcing, and so on.

The Copenhagen Diagnosis www.copenhagendiagnosis.com
A link to text originally drafted on climate science for the Intergovernmental Panel on Climate Change's (IPCC) fourth assessment report and a more recent updated version. The ski-slope diagram referred in the text originated in this report.

And Then There's Physics https://andthentheresphysics.word-press.com
A useful website by a UK scientist Ken Rice at the University of Edinburgh, with a series of useful blogs and articles containing data, figures, and discussions on science.

FURTHER RESOURCES

Skeptical Science https://skepticalscience.com
A website that addresses skeptical questions about climate science.

NOAA PMEL Carbon Program https://pmel.noaa.gov/co2
A website with data and information on ocean carbon update.
Carbonbrief.org https://www.carbonbrief.org/
A website with up-to-date articles on science, energy, and policy
related to carbon issues.
Climatecentral.org https://sealevel.climatecentral.org
A website providing graphical data on sea-level rise and its impli-
cations for human population, based on peer-reviewed scientific
information.

Articles

*On the Temperatures of the Terrestrial Sphere and
Interplanetary Space*
Jean-Baptiste Joseph Fourier, translated by R. T. Pierrehumbert

*On the Influence of Carbonic Acid in the Air upon the
Temperature of the Ground*
Svante Arrhenius, Philosophical Magazine and Journal of Science
Series 5, Volume 41, 1896, pp 237-276.

*Joseph Fourier, the 'greenhouse effect', and the quest for a uni-
versal theory of terrestrial temperatures*
James R. Fleming, Endeavour, Vol. 23(2) 1999

*Arrhenius' 1896 Model of the Greenhouse Effect in Context
Author(s): Elisabeth Crawford Source*: Ambio, Vol. 26, No. 1,
Arrhenius and the Greenhouse Gases (Feb., 1997), pp. 6–11
Published by: Allen Press on behalf of Royal Swedish Academy

of Sciences Stable URL: http://www.jstor.org/stable/4314543 Accessed: 02/10/2009 17:51

Sustainable carbon emissions: The geologic perspective, Donald J. DePaolo, MRS Energy & Sustainability: A Review Journal page 1 of 16, Materials Research Society, 2015

Early onset of significant local warming in low latitude countries, I. Mahlstein, R. Knutti, S. Solomon and R.W. Portmann, Environ. Res. Lett. 6 (2011) 034009

Physical and biogeochemical modulation of ocean acidification in the central North Pacific, J.E. Dore et al., Proc. Natl. Acad. Sci. 106, 12235 (2009)

Climate tipping points — too risky to bet against, Timothy M. Lenton, Johan Rockströ, Owen Gaffn, Stefan Rahmsto, Katherine Richardson, Will Steffen, and Hans Joachim Schellnhuber, Nature, 575 , 592 27 Nov. 2019 · Correction, 09 April 2020

Climate change now detectable from any single day of weather at global scale,
Sebastian Sippel, Nicolai Meinshausen, Erich M. Fischer, Enikő Székely and Reto Knutti, Nature Climate Change, Vol 10, 35 (2020):

History of the Greenland Ice Sheet: Paleoclimatic insights
Richard B. Alley, J.T. Andrews, J. Brigham-Grette, G.K.C. Clarke, K.M. Cuffey, J.J. Fitzpatrick, S. Funder, S.J. Marshall, G.H. Miller, J.X. Mitrovica, D.R. Muhs, B.L. Otto-Bliesner, L. Polyak, J.W.C. White, Quaternary Science Reviews (2010), doi:10.1016/j.quascirev.2010.02.007

Can we still avoid Dangerous Human-Made Climate Change, James Hansen, Feb 10, 2006 (online http://www. columbia.edu/~jeh1/2006/NewSchool_20060210.pdf) and published in Social Research: An International Quarterly, Volume 73, Number 3, Fall 2006, pp. 949–971

Sea-level fluctuations during the last glacial cycle, M. Siddall, E. J. Rohling, A. Almogi-Labin, Ch. Hemleben, D. Meischner, I. Schmelzer, and D. A. Smeed, Nature, 423, 853 (2003)

Ice-Sheet and Sea-Level Changes, Richard B. Alley, Peter U. Clark, Philippe Huybrechts, Ian Joughin, Science, 310, 456 (2005)

Ongoing climate change following a complete cessation of carbon dioxide emissions Nathan P. Gillett1, Vivek K. Arora, Kirsten Zickfeld, Shawn J. Marshall, and William J. Merryfield, Nature Geoscience, 4, 83 (2011)

The Abundance of O18 in Atmospheric Water and Water Vapour, W. Dansgaard (1953) Tellus, 5:4, 461-469, DOI: 10.3402/tellusa.v5i4.8697

A Reconciled Estimate of Ice-Sheet Mass Balance, A Shepherd et al., Science, 338, 1183 (2012)

The Amazing Case of "Back-Radiation", https://scienceofdoom.com/2010/07/17/the-amazing-case-of-back-radiation/

Mekong delta much lower than previously assumed in sea-level rise impact assessments, P.S.J. Minderhoud, L. Coumou, G. Erkens, H. Middelkoop, E. Stouthamer, Nature Communications 10:3847 (2019)

New elevation data triple estimates of global vulnerability to sea-level rise and coastal flooding, S. A. Kulp, B. H. Strauss, Nature Communications 10:4844 (2019)

FIGURES AND PHOTO CREDITS

MOST OF THE FIGURES AND PHOTOGRAPHS I HAVE INCLUDED IN this book to help clarify concepts and illustrate existing data, beyond those I created myself, reside in the public domain. I thank all those individuals and organizations who have produced these graphical aids for the benefit of public education and understanding. When figures are not in the public domain, I thank the creators for permission to use their work in this book. The source of images used here is given below, along with attribution where appropriate.

1   Mekong river. Photo credit: Lawrence Krauss
2   Tide, figure 1—Lawrence Krauss
3   Tide, figure 2—Lawrence Krauss
4   Tide, figure 3—Lawrence Krauss
5   Keeling Curve—Scripps Institution of Oceanography, SIO
6   Antarctic Ice Core Sites—US ITASE, http://www2.umaine. edu/USITASE/figures/fig6.html
7   Keeling Curve to 1700—Scripps Inst. of Oceanography, SIO
8   Keeling Curve to 10K back—Scripps Inst. of Oceanography, SIO
9   Keeling Curve to 800K back—Scripps Inst. of Oceanography, SIO
10 Keeling curve NOAA Climate.gov—Data from NCEI
11 Global Fossil CO_2 Emissions 2018—Future Earth and Global Carbon Project 2018. Supplemental data of Global Carbon Budget 2018 (Version 1.0) [Data set]. DATA CDIAC/GCP/BP/USGS; https://doi.org/10.18160/gcp-2018

12 Keeling Curve back to year 1—https://ourworldindata.org/
    co2-and-other-greenhouse-gas-emissions Source, NOAA/
    ESRL (2018), *Dr. Pieter Tans, NOAA/GML (www.esrl.*
    noaa.gov/gmd/ccgg/trends/) and Dr. Ralph Keeling, SIO
    (scrippsco2.ucsd.edu/). https://www.esrl.noaa.gov/gmd/
    ccgg/trends/data.html

13 Box model version of global carbon cycle—courtesy of
    Donald J. DePaolo, LBL

14 Our World in Data based on Global Carbon Project;
    Carbon Dioxide Information Analysis Centre; BP;
    Maddison; UNWPP; https://ourworldindata.org/grapher/
    annual-co-emissions-by-region

15 Adapted from box model version of global carbon cycle—
    Donald J. DePaolo, LBL

16 Figure showing solar radiance on Earth—Lawrence Krauss

17 Solar and Earth radiation spectrum, as ideal black bodies—
    Lawrence Krauss

18 Radiation balance between Earth and atmosphere—
    Lawrence Krauss

19 Detailed Earth-Atmosphere energy balance—National
    Weather Service, NOAA https://www.weather.gov/
    jetstream/energy

20 Downward Longwave Radiation, Billings, OK—Lawrence
    Krauss, created from World Radiation Monitoring Center
    BRSN data station no. 28, Latitude: 36.605000, Longitude:
    -97.516000, Elevation: 317.0, Data source: Sisterson,
    Douglas L (2009): Basic measurements of radiation at
    station Billings (1993-10). Argonne National Laboratory,
    PANGAEA, https://doi.org/10.1594/PANGAEA.723372

21 Net radiation flowing into the Earth—NASA Earth
    Observatory

22 Schematic of O_2 and CO_2 molecules—Lawrence Krauss

FIGURES AND PHOTO CREDITS

23 Transparency of Atmosphere—Arrhenius 1896 paper, Phil. Mag. & J. Science

24 Variation of Temperature—Arrhenius 1896 paper, Phil. Mag. & J. Science

25 Absorption spectrum of H2O and CO_2—NASA Earth Observatory

26 Schematic of CO_2 absorption—Lawrence Krauss

27 Schematic of CO_2 absorption—Lawrence Krauss

28 Schematic of CO_2 absorption—Lawrence Krauss

29 Measurements of downward and upward radiation at Arctic ice sheet—Data D. Tobin, Figure courtesy Grant Petty

30 Model atmosphere calculation of upward irradiance—Wikimedia Commons, Incredio, https://creativecommons.org/licenses/by-sa/3.0/deed.en

31 Schematic of effect of radiative forcing—Lawrence Krauss

32 Reproduction of H. Koch CO_2 Tube experiment with predictions, R. T. Pierrehumbert, realclimate.org

33 History of Global surface temperature since 1880—Climate.gov (2020)

34 Radiative forcing since 1980—EPA.gov, data from NOAA 2016

35 Table of Radiative Forcing—IPCC Fifth Assessment Report (AR5) 2013 Summary for Policymakers. In: *Climate Change 2013: The Physical Science Basis. Contribution of Working Group I to the Fifth Assessment Report of the Intergovernmental Panel on Climate Change* [Stocker, T.F., D. Qin, G.-K. Plattner, M. Tignor, S.K. Allen, J. Boschung, A. Nauels, Y. Xia, V. Bex and P.M. Midgley (eds.)]. Cambridge University Press, Cambridge, United Kingdom and New York, NY, USA.

36 Vostok Ice Core record—Lawrence Krauss, based on data from J. R. Petit et al., Nature 399, 429 (1999)

37 Dome C Ice Core record, figure courtesy of J. Hansen, based on data from D. Luthu et al., Nature, 453, 379 (2008)

38 Paleoclimate Temperature Change—courtesy of J. Hansen (2006)

39 Long term Atmospheric Carbon—S. Solomon et al. PNAS, 106, 1704 (2009)

40 Model Long term Global Temperature—S. Solomon et al. PNAS, 106, 1704 (2009)

41 Temperature variation across the globe—Ongoing climate change following a complete cessation of carbon dioxide emissions, N. Gillett et al, Nature Geoscience 4, 83 (2011) Reprinted by permission from Springer Nature

42 Global Mean sea level rise—UCAR R.S. Nerem, Nat. Center for Atmospheric Research Staff, 2016, https://climatedataguide.ucar.edu/climate-data/global-mean-sea-level-topex-jason-altimetry

43 Model long term sea level rise—S. Solomon et al. (2009)

44 Precipitation variation across the globe, Long Term Model—Gillett et al. (2011) Reprinted by permission from Springer Nature

45 Precipitation reductions during dry season in 3000 -Solomon et al. (2009)

46 Ski slope curve of Maximum Emissions reduction—From German advisory council on global change Factsheet, Climate Change: Why 2°C?, WBGU, (2009)

47 Emissions reduction needed for 2 degrees temp rise, Our world in Data https://ourworldindata.org/grapher/co2-mitigation-2c

48 Global greenhouse gas emissions scenarios, https://ourworldindata.org/future-emissions Licensed under CC-BY by Hannah Ritchie and Max Roser

49 Global Carbon emissions forecast—Courtesy of Donald J. DePaolo, LBL

50 Greenland ice mass loss—I. Velicogna, Geophys. Res. Lett, Oct 13, 2009 https://agupubs.onlinelibrary.wiley.com/doi/full/10.1029/2009GL040222, https://doi.org/10.1029/2009GL040222

51 Greenland ice mass loss—NOAA Arctic Report Card 2-19. M. Tedesco et al.

52 Greenland Melt Days—NOAA Arctic Report Card 2007, 2012, 2019

53 Standardized melt index—NOAA Climate.gov from 2012 Arctic Report Card

54 Future Greenland Ice Sheet—Ice Sheets and Sea-Level Changes, R.B. Alley et al., Science 310, 456 (2005). Reprinted with permission from AAAS

55 Iceberg off West Antarctic Peninsula—Lawrence Krauss

56 Larsen C rift, NASA, Earth Observatory; NASA photographs by John Sonntag

57 Temperature change in Antarctica, NASA EARTH OBSERVATORY, 2009 Based on data from Eric J. Steig et al., *Nature* 2009 Jan 22. Image courtesy Trent Schindler, NASA Goddard Space Flight Center Scientific Visualization Studio

58 Satellite image of Amundsen Sea—NASA/GSFC/SVS https://www.nasa.gov/jpl/news/antarctic-ice-sheet-20140512/

59 Schematics of melting of floating ice sheets—Lawrence Krauss

60 Antarctic & Greenland Ice mass loss—Figure Courtesy of A. Shepherd, based on data accumulated by the IMBIE team, as described in *Nature* | Vol 579 | 12 March 2020 | 233, 14 June 2018 Vol 558, Nature 219 and Science, 338 6111:1183-9

61 Paleoclimate sea level reconstruction by Robert A. Rhode, data, Fleming et al., 1998, Fleming 2000, Milne et al 2005 https://commons.wikimedia.org/wiki/File:Post-Glacial_Sea_Level.png

62 Vostok Ice Core record—Lawrence Krauss

63 Sea Level rise from Red Sea data—courtesy of J. Hansen (2006)

64 Single Day Climate change temperature anomalies—Sippel, S. *et al.* Climate change now detectable from any single day of weather at global scale. *Nature Climate Change* 10, 35–41 (2020). Reprinted by permission from Springer Nature.

65 Signal vs. Noise for an oscillating sinusoidal wave—Lawrence M. Krauss

66 Signals vs. Noise, global models—Mahlstein et al., Env. Res. Lett. (2011)

67 Evidence of Global warming signal—Mahlstein et al., Env. Res. Lett. (2011)

68 Per person regional carbon emission—S. Solomon

69 Warming to variability ratio, per country—from Mahlstein et al. (2011)

70 pH measurements and keeling curve—www.pmel.noaa.gov/co2, from Feely, R.A., R. Wanninkhof, B.R. Carter, P. Landschützer, A.J. Sutton, and J.A. Triñanes (2018): Global ocean carbon cycle. In State of the Climate in 2017, Global Oceans. Bull. Am. Meteorol. Soc., 99(8), S96–S100, doi: 10.1175/2018BAMSStateoftheClimate.1

71 Ocean acidity and Pteropods dissolution—NOAA.gov NOAA Environmental Visualization Laboratory

72 Mekong elevation data—P.S.J. Minderhoud et al., *Nature Comm.* (Aug 28, 2019)

73 South Vietnam elevation data—P.S.J. Minderhoud et al. (2019)

74 Predicted flooding 2100 South Vietnam—P.S.J. Minderhoud et al. (2019)

75 South Vietnam under High Tide 2050 Climate central—https://sealevel.climatecentral.org Based on S. A. Kulp, ,B. H. Strauss, *Nature Communications* (2019)

INDEX

INDEX

INDEX

ABOUT THE AUTHOR

LAWRENCE M. KRAUSS IS AN INTERNATIONALLY KNOWN THEOretical physicist and the author of hundreds of articles and numerous popular books including NYT bestsellers, The Physics of Star Trek and A Universe from Nothing. He received his PhD from MIT and then moved to the Harvard Society of Fellows. Following eight years as a professor at Yale University, he was appointed as a full professor with an endowed chair while still in his thirties. He has made significant contributions to our understanding of the origin and evolution of the Universe and has received numerous national and international awards for his research and writing. He is currently President of The Origins Project Foundation and host of The Origins Podcast with Lawrence Krauss. Between 2006 and 2018, he was Chair of the Board of Sponsors of the Bulletin of the Atomic Scientists. He appears regularly on radio, television, and film.